QUANTUM PHYSICS
FOR BEGINNERS

© Copyright 2020 by Roger Smith.
All right reserved.

The work contained herein has been produced with the intent to provide relevant knowledge and information on the topic on the topic described in the title for entertainment purposes only. While the author has gone to every extent to furnish up to date and true information, no claims can be made as to its accuracy or validity as the author has made no claims to be an expert on this topic.

Notwithstanding, the reader is asked to do their own research and consult any subject matter experts they deem necessary to ensure the quality and accuracy of the material presented herein. This statement is legally binding as deemed by the Committee of Publishers Association and the American Bar Association for the territory of the United States. Other jurisdictions may apply their own legal statutes. Any reproduction, transmission, or copying of this material contained in this work without the express written consent of the copyright holder shall be deemed as a copyright violation as per the current legislation in force on the date of publishing and subsequent time thereafter. All additional works derived from this material may be claimed by the holder of this copyright.

The data, depictions, events, descriptions, and all other information forthwith are true, fair, and accurate unless the work is expressly described as a work of fiction. Regardless of the nature of this work, the Publisher is exempt from any responsibility of actions taken by the reader in conjunction with this work.

The Publisher acknowledges that the reader acts of their own accord and releases the author and Publisher of any responsibility for the observance of tips, advice, counsel, strategies, and techniques that may be offered in this volume.

TABLE OF CONTENTS

Introduction 7
Chapter 1 What is Quantum Physics? 10
The Grandfather of Atomic Study 12
Avogadro and His Gases 14
Constructing the Atom 18
Radioactivity, Isotopes, and Pioneering Research 22
Conserving Matter, Energy, and Mass 29
Chapter 2 Particles, Waves, and the de Broglie Equation 39
Particle Theory 40
Wave Theory 48
de Broglie's Hypothesis of Particles and Waves 53
de Broglie's Equation 56
Chapter 3 The Bohr Model, the Schrödinger Equation, and the impact of de Broglie's Hypothesis 59
Rethinking Atomic Structure and Particle Wave Duality 61
Relating Schrödinger's Equation to de Broglie's Equation 65
Adapting the Fundamentals as Knowledge Evolved ... 67
Chapter 4 The Planck Constant 70
Max Planck and His Early Work 71

Black Bodies and the Electromagnetic Spectrum 74
Planck's Law and Development of the Constant 77
Measurement and behavior Planck's constant in action 80
Development and use of Planck's reduced Constant another use for Planck 83

Chapter 5 Heisenberg's Uncertainty Principle 87

Heisenberg's Beginnings in Physics 88
The Development of the Uncertainty Principle 91
Mathematics uncertainty and the Planck Constant in Action 96
Separating uncertainty from the observer effect 98
Corrections, rebuttals, and adaptations of the Uncertainty Principle 100

Chapter 6 Einstein and his Foundational Physics 103

Early Life and Work 104
Brownian Motion 107
The Photoelectric Effect 109
General relativity, Special relativity, and Mass Equivalency 115
Later Years and Lasting Impacts of Einstein's Work 131

Chapter 7 A glimpse into the future of

Quantum study **137**
Quantum Mechanics in the 140
21st Century 140
Quantum Physics in the 143
21st Century 143
Appendix A: Timeline of Major Breakthroughs in Early Quantum Physics **152**
Appendix B: Formulas and Equations **155**
Conclusion **158**

Introduction

Welcome to Quantum Physics for Beginners! Just the phrase "quantum physics" can strike fear into the heart of the most science-savvy, but this book is here to demystify the basic principles that form the fundamentals of quantum physics and give you a firm understanding from which you can continue to grow your knowledge.

Although you may not think about it, you use quantum physics daily. Every time you flick on a lightbulb or get in your vehicle to drive somewhere, quantum physics played a role in bringing you that technology. Simply put, quantum physics is the study of how the tiniest pieces of our universe work and interact. The field attracts the brightest minds in both science and mathematics, who have collectively spent the better part of the last century finding new ways to describe, explain, and manipulate atoms and molecules.

The impact of quantum physics on the modern world is not to be ignored. Without quantum physics, space travel and exploration would be impossible. The satellites that orbit the Earth would never have been built. We wouldn't have the advanced medical diagnostic equipment that so many health care providers depend on for imaging and treatment. The computer chips that power our cell phones, the coolant that runs through our refrigerators and air conditioners, and the burning of the sun itself are all powered by quantum physics.

Quantum Physics for Beginners will help you learn the history behind the science, including Bohr's earliest models of the atom and how others used his work to build their own theorems. We'll look at the relationship between Bohr's model and the de Broglie equation, and the advancement of particle and wave theory brought about by this development.

Moving forward from these original hypotheses, we'll enter into a discussion about one of quantum physics underlying rules, the Planck Constant, and its counterpart, the Heisenberg Uncertainty Principle.

The book will finish up with an in-depth examination of the work of perhaps the world's most famous physicist,

Albert Einstein. Einstein's theorems and subsequent equations formed the basis of modern quantum physics; he brought together and coalesced the building blocks of study that are used today and will be used well into the future. Einstein's seminal works, including his Theory of Relativity, are the basis for nearly all quantum physics study and advancement in the 21st century. Einstein was able to take the knowledge accumulated by others before him, create theories that tied together early principles, a bridge the knowledge gap between the history and the future of quantum physics.

Quantum Physics for Beginners aims to give you some fundamental insight into the fantastic world of science and math that rules how the universe works. With easy-to-read explanations and definitions, this book will break down the equations and the history behind them and look at the extraordinary men who discovered the universe's smallest fragments and attempted to harness them for science. If you're ready to dive into the world of quantum physics, let's get started.

The wonders of the universe await you, let's go!

Chapter 1

What is Quantum Physics?

The study of quantum physics is one of science's newer disciplines, but it has its roots in centuries of accumulated knowledge. Physics itself is a broad scientific field, encompassing the study of nature, matter, and energy. It includes research and analysis of how all matter acts and reacts, such as through light, sound, kinetic energy, magnetism, and the behavior of the atom. Quantum physics seeks to answer questions about matter and energy on its smallest, most fundamental levels and broadest, most universal levels. Looking at the history of the factors at play in modern quantum physics gives insight into how these concepts shape the field of study we see today.

Quantum physics grew from the humble beginnings of classical physics, which has been studied since the Sumerians first put chisel to tablet. The simple machines that we all learned about in grade school are examples of classical physics in action. And we're sure that everyone's heard the story of Archimedes discovering volume and displacement in his ancient Grecian bathtub. The truth is, physics is all around us, all the time. Gravity is what's keeping you from floating away right now. Physics keep your car running and, on the road, and your headlights illuminating your way. Without physics, we wouldn't be able to enjoy the lifestyle that we do in the modern age. Since we're here to talk about quantum physics, let's get things started with a look at the origins of the discipline, beginning with the first atomic theory.

The Grandfather of Atomic Study

It's the atom itself that forms the basis for the specialized field of quantum physics.

Although the atom was first described in 400 B.C. by a Greek philosopher named Democritus, it wasn't until 1803 that the first scientific atomic theory was developed by British chemist John Dalton.

Dalton was a pioneer in predictive meteorology and the study of genetic color blindness before moving onto atomic chemistry. He released a theorem in 1808, which detailed what he considered to be the five properties of the atom.

1- Atoms cannot be destroyed or divided

2- All the atoms in a single element are identical

3- Atoms of different elements have different properties and weights

4- The atoms of different elements can be combined in simple numbers to form molecules (Dalton used the word "*compounds*")

5- Atoms can be neither created nor destroyed; all matter will break down in recoverable, unchanged atoms.

Using these principles, Dalton would also create the first rudimentary periodic table. It contained only six elements- hydrogen, oxygen, nitrogen, carbon, sulfur, and phosphorus- but showed the relative weights of an atom of each element based on hydrogen having a value of one (1). Dalton gave the scientific community a firm basis on which to build the field we now know as quantum physics. In fact, very little has changed in the over two centuries since Dalton first published his atomic theory in a booklet called *A New System of Chemical Philosophy*.

The only significant edit in his theory in the two-plus centuries since its publication is that we now know that the atom is not the smallest unit of matter; an atom's individual components can also be seen and measured. We also now know that an atom can be split, and we have the technology to do so.

Avogadro and His Gases

Using the work of Dalton as his basis, Italian scientist Amedeo Avogadro began his groundbreaking study of the behavior of gases. Avogadro believed there might be a flaw in one of Dalton's theories on this topic. While there was no flaw in Dalton's physical work, there was a small mistake in his interpretation of how water absorbed carbon dioxide, nitrogen, and other various gases. Dalton believed that the water behaved differently given the concentration of the gases. In contrast, Avogadro would go on to prove that it was instead the atomic weight of the gases that created the differing reactions.

Avogadro's most prominent legacy in the field of quantum physics is his namesake number, seen here:

$$6.02214076 \times 10^{23} = 1 \text{ mole}$$

This equation represents the number of particles (atoms, molecules, ions, etc.) that are contained within a substance held at a specific volume, pressure, and temperature. This unit is now known as a mole and is recognized as an SI Unit with the symbol mol. Avogadro theorized and subsequently proved that this is a universal truth that governs all gases and that an equal volume of any gas at the same temperature and pressure will contain this number of particles, regardless of atomic weight. A beautiful thing about Avogadro's equation is that it can be used to convert atoms to moles and moles to atoms, based on what knowledge the scientist already possesses. This is because the molar weight of a substance and the atomic weight of the substance are the same. For example:

- Water molecules are made up of two hydrogen atoms and one oxygen atom
- The combined molecular weight of a water molecule is 18.015 amu (atomic mass units)
- Therefore, a mole of water weighs 18.015 grams, expressed as g/mol.

Being able to calculate the atomic weights and convert back and forth from mass to molar units makes it that much easier for scientists to work with large numbers and comprehend the vast number of atoms that make up every known substance.

Let's look at a calculation in which we know the atomic weight but need to calculate the number of atoms within a known sample of carbon, which has an atomic weight of 12 amu.

Carbon is regularly used as the standard against which all other atomic weights are measured because this is the substance upon which Avogadro built his equation. Therefore:

- 12 grams of carbon-12 have an atomic quantity equal to 1 mol (6.022×10^{23})

- To calculate the molar weight or molar quantity of another substance, simply plug in the number which you do know and the variables which you don't; these equations look like this:

If you know the number of moles (x) but need to calculate the number of atoms (y), use this equation:

$$\frac{x \; moles \cdot 6.022 \times 10^{23}}{1 \; mole} = y \; atoms$$

Reversing the calculation above, it is possible to convert a number of atoms to a molar quantity by dividing it by Avogadro's number.

If you know the number of atoms (x) but need to calculate the number of moles (y), use this form of the equation:

$$\frac{x \text{ atoms}}{6.022\times10^{23} \text{ atoms}} = y \text{ moles}$$
$$\frac{}{1 \text{mole}}$$

This can be written without a fraction in the denominator by multiplying the number of atoms by the reciprocal of Avogadro's number:

$$x \text{ atoms} \cdot \frac{1 \text{ mole}}{6.022\times10^{23}} = y \text{ moles}$$

Because it is so useful in calculating atomic content and molar weights, Avogadro's number is often referred to as Avogadro's Constant. Avogadro's Constant is used to calculate the volume of a substance within any measurable space.

It's especially useful in letting scientists communicate large numbers of particles with one SI Unit.

Constructing the Atom

Without working knowledge of the structure of the atom, how would it be possible to study its behavior and determine its properties? Simply put, it wouldn't be, and so it's essential to recognize the work of the scientists who worked to devise the first models of an atom as we understand them today. These early models weren't perfect but gave the researchers who came later a better understanding of how these tiny particles of matter work and interact.

One of the earliest models of the atom was created by British physicist J.J. Thomson in 1904, known as the 'plum pudding' model. Thomson is credited with discovering the negatively-charger atomic sub-particle now called the electron. Thomson realized that for an atom to be held together, there must also be a counteracting positive charge. Named for the popular British dessert of bread pudding with raisins, the plum pudding model of the atom showed a field of positivity (the pudding) embedded with negative electrons (the raisins). Thomson was on the right track, but the model of the atom wasn't quite there yet.

The next advancement in the working model of an atom was through the research of Ernest Rutherford and his students Hans Geiger and Ernest Marsden, in 1911. The Geiger-Marsden experiments involved bombarding thin gold foil with alpha rays. The students and Rutherford observed that only about 90% of the rays went through the foil. The other percentage was deflected, leading the scientists to believe that something was causing the deflection. In turn, this observation steered them to the hypothesis that each atom actually had a center, or nucleus, that was capable of turning away the stream of alpha rays. The resulting model was the first cloud representation of the atom, one with a nucleus with electrons floating in regular orbits, rather than randomly bouncing around, as the plum pudding model had portrayed.

In 1913, working with Danish physicist Niels Bohr, the model was updated slightly to recognize that the nucleus of the atom was made of the subatomic particle now known as the proton. The proton and the electron work together to keep the atom at a neutral electrical charge. The Rutherford-Bohr model is more commonly simply called the Bohr model. It's this representation that the vast majority of people, from the youngest of elementary school science students to the most advanced theoretical physicists, use and are familiar with today.

It wasn't until physicist and former Rutherford student James Chadwick discovered the neutron in 1932 that the full picture of the atom came into focus. While the research was rapidly advancing in the field of radioactivity (more on that in a bit), scientists were finding it difficult to reconcile the atomic weights of elements based solely upon the presence of protons and electrons. While the number of protons in an atom defines its atomic number, the mass of the nucleus determines its atomic weight. So, *where was the differential coming from?* Chadwick theorized that there must be another particle in the nucleus affecting the atomic weight, but not the electrical charge of the atom.

Using this hypothesis, Chadwick conducted a series of experiments using alpha and gamma radiation to prove his theories. The results showed the exposure of a new subatomic particle, the neutron.

Neutrons are intermixed with protons in the nucleus of the atom and solved the puzzle of why atomic weights were not equal to atomic number. This development in the understanding of the construction of the atom earned Chadwick as Nobel Prize for Physics in 1935 and forever changed the face of quantum physics.

Radioactivity, Isotopes, and Pioneering Research

The funny thing about the discovery of subatomic particles and the first accurate models of the atom is that these advancements came late in the game in the initial wave of developments in quantum physics.

The creation of these accurate models, however, gave the scientists who followed the ability to look back at the work of their predecessors and use their research to grow the field of quantum physics by leaps and bounds.

The work of the earliest particle physicists is nothing to be ignored. One of the most important discoveries of the late 19th century is that of radioactivity.

French scientist Henri Becquerel, who would also work with Pierre and Marie Curie, was experimenting with phosphorescent minerals when he stumbled upon what would be the first recorded instance of spontaneous radioactivity while studying uranium salts. Spurred by the discovery of X-rays by his colleague Wilhelm Röntgen early in 1896, Becquerel speculated that uranium salts could function in much the same way and thought he could harness the power of their phosphorescence by exposing them to bright light like sunlike.

What Becquerel would soon discover is that he didn't need a light source to activate the phosphorescence of the uranium salts. Coupled with research on thorium, as well as work on polonium and radium being conducted by the Curies, theories and proof of natural radioactivity would snowball in the years leading up to the turn of the 20th century. Ironically enough, a colleague of Becquerel's father, both of whom were also early physicists, had almost accidentally discovered radioactivity nearly forty years prior to the younger Becquerel's findings.

That scientist, Frenchman Abel Niépce de Saint-Victor, was researching photography and photosensitive processing materials when he observed that uranium-based chemicals could expose photographic plates before they had been subjected to light-processing.

If he had been curious enough to go one step further into why uranium had this effect on those photographic plates, he might have been the one to win the Nobel Prize for the discovery.

Scientists were able to harness the power of radioactivity even before they fully understood precisely what was causing this behavior. Along with the discovery of x-rays, early work using alpha, beta, and gamma rays was also underway. Researchers were beginning to understand the "how" of radioactivity and radiation, even if they didn't yet understand the "why." Using these rays was a huge advance in early quantum physics, and scientists like Becquerel, the Curies, Rutherford, and their students and cohorts were able to make significant advances in their experimentation through the use of rays.

What these pioneering physicists did understand is that every atom of every element is in a constant state of flux. The subatomic particles that make up an atom must move in order to produce the electromagnetic energy necessary to hold the atom together.

This movement produces waste energy, which is emitted in the form of radiation. Some elements are more stable than others, and therefore require little energy to maintain their structure. These elements have less radioactivity, which is the measure for how much energy, or radiation, an atom emits. Other elements, like radium, uranium, and thorium, are much less stable. They require a large amount of atomic energy to maintain their form. Consequently, these elements have a higher measure of radioactivity.

Using this knowledge, scientists were able to start concentrating this radiation. X-rays, of course, were discovered to be useful in exposing images hidden within solid objects when combined with photographic processing. It was Rutherford who classified alpha, beta, and gamma rays, allowing his students to use alpha radiation to complete the famous Geiger-Marsden experiments, which allowed them to hypothesize the true nature of the empty space inside an atom. Rutherford named and categorized the types of radiation based on their abilities to penetrate other solid substances. Alpha radiation is made up of larger, slower-moving particles. Beta radiation is faster and made of slightly smaller particles than alpha radiation. Gamma radiation consists of tiny, swiftly moving particles. Because gamma rays carry little to no electrical charge, they can penetrate most objects with ease, regardless of density or mass. There are, of course, other types of radiation; electromagnetic emissions like microwaves and infrared, ultraviolet, and visible light are also classified as radiation.

Radioactivity, the term coined by Becquerel and made famous by the Curies, is directly related to the nucleus of the atom and its natural breakdown.

Once the discovery of radioactivity was made, it wasn't much of a leap to discover precisely what caused it and what set some atoms apart from others.

It would be another Rutherford colleague, Frederick Soddy, who would first hint at the existence of isotopes-variations on the subatomic makeup of atoms of the same elements. As it turns out, the nucleus of an atom, which consists of protons and neutrons, can have a variable number of neutrons, which affects the stability of the nucleus. The result of this variable number of neutrons are isotopes, which are defined as different species of the atoms which make up and element.

In elements with high radioactive values, like uranium or thorium, isotopes are often unstable and frequently shedding neutrons leading to rapid change and breakdown. Becquerel and Marie and Pierre Curie, whose pioneering research into these types of highly radioactive substances, would also be among the first to recognize that the heavy gamma radiation emitted was also the cause of terrible, irreversible cell damage, which we know now to be radiation poisoning. Becquerel died from the complications of extensive burns, later realized to have been triggered by his unprotected handling of uranium salts. Perhaps Pierre Curie would have also been taken by radiation poisoning, had he not been killed in a carriage accident in 1906. His life and scientific achievement were cut tragically short, but Marie carried on their work with the help of their daughter Irene until her own death from radiation-related leukemia in 1934.

She remains the only woman to have won two Nobel prizes; the first in Physics was for the Curies' work alongside Becquerel in 1903, and the second in Chemistry for her discovery of polonium and radium.

All elements have a measure of radioactivity.

The research published by Becquerel, Rutherford, and the Curies led to a far greater understanding of the nature of the atom, how to measure radioactive decay, and how to use radiation safely and purposefully.

The discovery of the existence of isotopes is what gave the world mobile x-ray machines, carbon-12 dating, and eventually, nuclear power and nuclear weapons.

This is all thanks to isotopes and their decay patterns.

If you'll recall John Dalton's properties of an atom, he states explicitly that atoms cannot be created or destroyed. While we now have the technology to prove Dalton wrong, he was also correct on a critical part of that property. It's matter itself that cannot be created or destroyed, as demonstrated in Antoine Lavoisier's 1789 Law of the Conservation of Mass.

Let's take a more in-depth look at what happens to matter during a chemical process or radioactive decay.

Conserving Matter, Energy, and Mass

There are laws that govern everything we know about physical matter and energy, and these universal laws form both the backbone and the limiting factors of physics throughout the history of the discipline.

The Law of the Conservation of Matter, also known as Conservation of Mass, states that no matter can be created or destroyed, only changed into another form. The Law of Conservation of Energy states that energy cannot be created or destroyed, only transferred or transformed. Together, these two laws tell us everything we need to know about balancing chemical and physical reactions, including radioactivity. These laws help us understand that everything that exists and moves in the physical world is created from something else; nothing is conjured out of nothing. The raw materials that make up all things are the tiniest subatomic particles and the energy they emit.

When you want to study the more advanced principles of quantum physics, it's crucial to keep these laws in the back of your mind. Before the conservation laws were hypothesized, tested, and cemented, alchemy was a popular practice, and surprisingly, alchemists had the right idea, even if they never did turn lead to gold.

It is possible to transform one element into another, but when Soddy was developing the research that led to the discovery of isotopes, he also was integral in creating the Law of Radioactive Displacement. This law proved that an atom in active radioactive decay could transform into another element and that the new element is determined by whether the original element is emitting either alpha or beta radiation. Working primarily with thorium (an isotope of radium), Soddy found that an atom that lost neutrons through alpha decay would transmute to become an element two spaces to the left on the periodic table, and those that lost neutrons through beta decay would transmute into an element one space to the right on the table.

Polish-American physicist Kazimierz Fajans, working in Rutherford's laboratory in Manchester, England, also independently developed the same hypothesis while researching the behavior of uranium; because of this, the Law of Radioactive Displacement is credited to both men. Fajans is also credited with groundbreaking research into the half-life values of uranium.

A half-life is the amount of time it takes for an atom to break down to half its original mass.

Therefore, after one half-life, there will be 50% left; after two half-lives, there will be 25% left, three will leave 12.5%, and so on.

Highly radioactive elements will have much shorter half-lives than stable elements. But if we're talking about both decay and conservation of matter, *where does the rest of the atom's original mass go?*

Let's look at an example that can help you clarify how you think about the conservation of matter.

Imagine you're going to have a campfire.

You pile up your wood, light a match, and your fire blazes up.

After a couple of hours, you're all out of fuel, and you're left with a heap of ashes where your pile of wood once was. The wood is completely gone, but if matter cannot be created or destroyed, *where did it go?*

The pile of ashes isn't nearly comparable in volume to the logs that you started with, so matter must have been destroyed, right? Wrong- it's just been transmuted. Think about the substances that make up a log of firewood; it contains organic compounds like cellulose, elemental nutrients, and water.

When these components are surrounded by airflow, which contains oxygen, nitrogen, and other atmospheric gases, and the chemical process of ignition and fire is applied, several things start to happen.

One of the first physical things you might notice when you light a campfire is that you'll hear sizzling and see steam. The heat of the fire is changing the phase of the water molecules contained in your firewood, and the liquids are becoming gases. The matter that was water initially contained within the log is the same amount of matter- it's just been released into the atmosphere.

As your fire burns, you'll begin to see more changes. The elements that make up the cellulose structures of the wood will start to break down into more basic molecules and eventually into its atomic components. The solid remnants will be present in the form of ash, whose mass will be much smaller than the original mass. That means that the rest of the elemental composition of your campfire will have been released into the atmosphere as the gases present in steam and smoke.

Your equation began as mostly solid, with a small amount of liquid content and the atmospheric gases present to spark the chemical reaction of fire. In the end, most of the matter will have been transformed into vaporous gases, leaving the solids elemental ashes behind. If you could trap and measure the gases and add it to the mass of the ashes, you would find them to be equal to the mass of wood and gases that you began with. You could think of the equation like this:

Firewood + atmospheric gases + catalyst (match) =

atmospheric gases + ashes

This is obviously a simplified look at the law in question, but it gives you a sound basis for thinking about matter as a constant.

The other law we're concerned about in this section is **The Law of Conservation of Energy**, so let's start to think about that in fundamental terms, as well.

The Law of Conservation of Energy is one of the oldest tenets of the study of physics. As a reminder, it states that energy cannot be created or destroyed but can change forms. There is a caveat to this because it can only be proven in a closed system where the energy cannot be acted upon by outside forces. Energy comes in several forms, which are expressed as either potential or kinetic. Potential energy is the energy that is being stored or built up within matter for future use.

In contrast, kinetic energy is the energy that matter uses when it is active or in motion. To visualize potential versus kinetic energy, think about a pendulum or a child on a swing. When the pendulum is at the highest point it can be, it is exhibiting potential energy. As soon as the pendulum begins to swing, it is exhibiting kinetic energy.

All energy can be categorized as follows:

Mechanical: This is the energy that is found in physical objects, and the total of mechanical energy is the kinetic plus the potential energy.

A moving object is using kinetic energy, making its potential energy zero.

A resting object shifts the equation in the other direction. An example of an object with a balance of kinetic and potential energy might be a car driving up a steep hill. The vehicle is moving, but not at its top speed, meaning that it is not using all of its potential energy.

Electromagnetic (Radiant): This is the form of energy that refers to anything that puts off electromagnetic waves or light, even non-visible spectrums like ultraviolet or infrared.

Electromagnetic energy can also be potential or kinetic, and this might be exhibited in something like a lightbulb and light switch.

The potential energy is being held in the closed circuit. When the switch is flipped, the circuit opens, allowing electricity to become kinetic, turning on the lightbulb, which further converts that electrical energy into light and heat. Microwaves, radio waves, and gamma rays are all also examples of electromagnetic energy.

Chemical: Chemical energy is the energy used or released during chemical processes or reactions.

A great example of the potential and kinetic energy of a chemical process is a stick of dynamite.

The dynamite exhibits potential energy before the application of the catalyst, in this case, fire, and when it explodes, it is displaying sudden and violent kinetic energy. It also converts some of its potential energy into sonic and thermal energy, which we'll cover in a moment. A less explosive example of chemical energy in real life might be a disposable battery powering a toy or a plant using its chlorophyll, water, atmospheric gases, and radiant energy to create glucose to feed itself and oxygen to emit.

Sonic: Sonic energy is, in fact, exactly what it 'sounds' like- this is the energy of sound waves.

Sound waves cannot exist in a vacuum; they must have another medium through which to travel, such as the air or water. Some good examples of sonic energy are the sound of your voice or music being played, or a sonic boom from a jet airplane.

Thermal: Thermal energy is the energy of heat. Heat is produced through a variety of chemical means, and it is distributed between systems through convection, conduction, or direct transfer. Simply put, heat is always trying to travel to where there is an absence of heat. Thermal energy is measured by finding the difference in the baseline temperature of the two systems.

Thermal energy is also essential in understanding how chemical and mechanical energy is transferred between systems.

Nuclear: Nuclear energy is what is created when the center part of the atom, the nucleus, is split apart through either mechanical or chemical means. It takes a great deal of force to break apart the nucleus, and that force is redirected into nuclear power, which can be harnessed for electricity and to power engines.

Nuclear energy, as most people know, can also be contained in weapons of mass destruction like bombs and warheads. One of the unfortunate byproducts of nuclear power is the resulting hazardous waste, which collects from the spent nuclear fuel and will take an exceptionally long time to cycle through its remaining half-lives to become inactive.

Gravitational: Gravitational energy is the force that keeps objects attracted to each. This best example of gravitational energy is the relationship between the sun and the planets or the Earth and the moon. Gravitational energy, or pull, is also what keeps up standing on the ground and not flying off into space.

Being able to understand gravitational energy is a fundamental part of being able to study and understand the physics of not only our planet but the solar system and beyond into deep space. Gravitational energy can explain astronomical phenomena that we can't see with our current technology.

If nothing else, quantum physics is about the relationship between the objects in our universe, from the tiniest subatomic particles seen with our most precise microscopes to the most massive stars and astronomical bodies beyond the scope of our strongest telescopes. Now that you've got a bit of history and a primer on energy and matter, it's time to move on to some of quantum physics' broad concepts and groundbreaking discoveries. We'll start by taking a closer look at the particles themselves and how they move through the world.

Chapter 2
Particles, Waves, and the de Broglie Equation

In the last chapter, we talked a lot about the properties of atoms and the nature of energy. The subatomic particles that make up atoms and the behavior of these atoms are the focus of quantum physics. In this chapter, we're going to dive into the way atoms and subatomic particles move, how this movement can be measured and affected, and the impact that subatomic movement has on the study of modern quantum physics.

Particle Theory

In order to understand particles, you must first know the principles of particle theory. The Particle Theory of Matter and the Particle Theory of Energy give us some universal truths about the tiniest pieces of our world. These theories are the absolutes on which quantum physics is based, so let's go over the tenets that make up the Particle Theories.

The Particle Theory of Matter consists of five simple statements:

All matter is made up of tiny particles, this first statement seems obvious, as even "nothing" is made of something. Still, without setting this baseline, it's impossible to build the rest of the particle theory and all other theories that make up quantum physics.

All individual substances are composed of their own type of matter, this principle is what allows scientists to categorize known elements and identify whether newly discovered substances are isotopes of previously recognized elements or potential new elements.

All particles are constantly in motion-, this movement is necessary to maintain atomic bonds. If the particles that make up an atom were to stop moving suddenly, the atom would fall apart.

The temperature directly affects how fast particles move- The warmer the particles, the quicker they move. You can see this in an experiment as simple as freezing some water, then letting it thaw and swirling it around a glass at room temperature, and then boiling it. Steam moves a lot faster than ice! Cold particles will slow down to conserve energy; warm particles have more energy to expend.

All particles exhibit attraction, the electrical charge carried by atoms and molecules means that all particles are looking to connect with other like-minded particles. It is these bonds that link atoms into elements and elements into compounds.

Knowing these five tenets of the Particle Theory of Matter will help you further understand our next fundamental of quantum physics, and that's the Particle Theory of Energy.

This theory is sometimes called the Kinetic Particle

Theory, and it explains and lays down the ground rules for the behaviors of particles at different temperatures and different states of matter.

Kinetic Particle Theory lists the following traits of matter in its differing states:

Solid: Matter is in a solid state when it is at a temperature that does not allow its particles to move freely. The particles in solids are arranged tightly together in a regular pattern, and they cannot move around; instead, they essentially vibrate in the space they are allotted. There is no room between the particles to allow for any other movement. Matter in a solid state holds its own shape due to the strength of the bonds between particles.

Liquid: Matter in a liquid state is matter that is at a temperature that allows the particles to spread out and take up more space. In a liquid state, the particles are no longer restricted to a tight pattern.

They have more space in between them and can flow more freely. These particles are also moving faster and more erratically than they were in a solid state.

Matter in a liquid state cannot hold its shape.

Rather, it takes the shape of its container.

Gas: Matter in a gaseous state is matter that has been heated to the point of boiling or evaporation.

The temperature is high enough to allow the particles to space themselves out in a random pattern and flow freely. If restricted, the particles will take the shape of their container, and if unrestricted, they will assimilate with the atmosphere. Particles in a gaseous state are the fastest moving and most erratic of all particles.

The temperature at which matter becomes solid, liquid, or gas is dependent on the substance.

Water (H_2O) becomes solid at 0* Celsius and becomes a gas at 100°C. Another common substance, isopropyl alcohol (C_3H_8O), doesn't freeze until it reaches -89°C but also becomes a gas at a lower temperature than water.

The boiling point of change for isopropyl is 80.4° C.

Every element and compound have its own set of temperatures that affect its state of matter.

It's essential to understand the phases of matter and the properties that accompany them, but it's also important to grasp the thermal expansion that underlies these changes of state. Thermal expansion does not increase the mass of the substance or its particles.

You will have the same amount of ice, water, and steam as you started with. What thermal expansion does increase is the volume of space between the particles in your substance. This expansion is what allows the particles to increase their speed and level of activity. Another item of note about the concept of thermal expansion is when a gas has reached its maximum entropy level but is contained and does not have room to keep expanding, the resulting physical force is pressure. That is why aerosol cans and other gas-filled devices shouldn't be subjected to excessive heat.

With no room left to move, the resulting rise in pressure can result in an explosion.

Phase changes occur through a few chemical processes. Gases are produced through either boiling or evaporation. Boiling, contrary to wide belief, is not entirely temperature dependent. The main difference between boiling and evaporation is that boiling requires active input from an energy source, such as a hot stove burner under a tea kettle. The water will begin to boil, and steam will be released. In contrast, evaporation would occur when an open saucepan is left out on a countertop. Over time, the most active (most excited) of the water molecules in the pan would *"escape"* from the surface into the atmosphere.

When matter moves from a gaseous to a liquid state, it is through the process of condensation literally, the particles are converting back from widespread erratic motion to a more condensed condition.

Condensation occurs when the temperature lowers to a point where the particles cannot maintain the amount of movement they exhibit when in a gaseous state. When the temperature lowers even further, liquids begin to freeze or solidify into their solid states.

The reverse of this process is solids melting into liquids. Both melting and boiling occur when a substance reaches its latent temperature or has been plied with the proper amount of latent heat. This is calculated by how much energy needs to be applied to the substance to begin the process of melting or boiling.

Instead of melting and boiling, you might sometimes see the terms "latent heat of fusion" (melting) and "latent heat of vaporization" (boiling).

There are two outliers to the Kinetic Particle Theory: The chemical process of sublimation and the existence of the fourth state of matter known as plasma. Sublimation is the change of matter directly from the solid phase to the gas phase, completely skipping the liquid phase. The best example of common sublimation is known as "*dry ice*," which is frozen carbon dioxide (CO_2). When it begins to melt, it doesn't have a liquid state; it becomes an evaporating gas immediately. This reaction can be hastened by placing the dry ice in room temperature water. Dry ice is useful for shipping and storing frozen items because of its extreme ability to emit cold, and it's also used for special effects like fog machines and Halloween decorations because of its ability to sublime.

Plasma is a bit of a strange concept to explain.

The so-called fourth state of matter occurs when particles are stripped of their electric charge, causing them to act in a completely erratic fashion.

Plasma is often considered to be a gas, but it doesn't behave in the same manner as a gas; plasma particles do not maintain even space between them, and they don't have a cohesive attraction.

These particles react readily to an electric charge, which is why common applications of plasma are neon or fluorescent lights and plasma televisions.

Now that we've laid the fundamentals of particle and energy theory, it's time to talk about exactly how these particle function and move. With these ground rules in mind, we're going to start taking a closer look at how these tiny particles make their way through the universe, one wave at a time.

Wave Theory

We've all been to the beach and have seen waves rolling into the shore or dropped a pebble into a puddle and watched the water ripple out from where the pebble broke the surface. We recognize this motion, but *have you ever stopped to think about how and why those waves exist? What about thinking about those waves on the tiniest scale? What makes a wave a wave?*

Waves are broadly classified into two categories, mechanical and electromagnetic, and we'll examine those classifications in a little while.

Before we talk about what makes waves different, let's first look at what makes waves similar. We already know that all particles are in motion all the time, even in a solid state. A wave occurs when those particles begin to move in an observable, measurable, often predictable manner. A wave cannot occur without the outside influence of other forces upon the natural movement of the particles. However, a wave is NOT a particle.

A wave is energy in motion and does not possess any mass.

Mechanical waves are a prime example of this. Mechanical waves are waves that move materials and sound, but they must have a medium through which to pass; mechanical waves cannot exist in a vacuum. Mechanical waves must also be created by an outside force- they do not happen spontaneously.

Using the illustration of a pebble dropped into a puddle, *what causes the waves?* The water, made up of H_2O molecules, is in a liquid state, meaning the particles are moving freely and taking up the shape of the puddle. Those molecules are still bonded to one another, and at the surface of the puddle, the energy they are using to stay connected results in surface tension.

When that surface tension is broken by the mass of the pebble, that energy (which, remember, cannot be created, or destroyed) must go somewhere, and so it is converted into wave energy. The height (or amplitude) of those waves will be determined by the velocity of the pebble, which in turn would be determined by the height and force at which it was dropped.

This will also affect the frequency of the waves, which is measured in how many waves pass a fixed point in a set amount of time, i.e., waves per second.

The second type of waves are electromagnetic waves, and this classification of waves includes all spectrums of light, x-rays, and gamma radiation.

Electromagnetic waves are made up of pure energy, and it is the amplitude and frequency of those waves that define what type of energy it is. Electromagnetic energy does not need a physical medium to travel through; as we know, light and other waves can travel through the vacuum of space.

Light energy is divided into a broad spectrum ranging from ultraviolet through visible light and ending with infrared. The scale is based on the wavelength (frequency) of this energy. Light energy falls in the middle of the electromagnetic wavelength scale, which begins with shortwave cosmic waves and culminates with longwave radio transmissions. The scale measures these waves in wavelengths and nanometers and goes from shortest wavelength to longest, like this:

Cosmic waves, Gamma rays, X-rays, Ultraviolet (UV) rays, Visible light spectrum (violet, indigo, blue, green, yellow, orange, red), Infrared, Microwave, Radar, Shortwave broadcast radio,

FM radio, Analog Television, AM radio, Longwave broadcast radio.

We don't think about it, but waves are all around us at any given time. Most of these rays are relatively harmless, but scientists learned lessons from the early days of working with x-rays and radiation that there can be harmful side effects of certain types of electromagnetic waves. It didn't take long to come up with simple, practical solutions to these exposure issues, which is why to this day, radiology technicians will wear and offer their patients lead-lined shields and aprons to prevent any tissue damage from unnecessary contact with x-rays. It's also why scientists and researchers wear protective suits and gloves when working with highly radioactive materials.

Waves are integral in the workings of the universe. Without waves, we wouldn't have radio, television, or microwave ovens. We wouldn't be able to communicate with people on the other side of the world, or even with our astronauts in space. Waves are responsible for every shade, tint, and hue of color that lights our world and the light that reaches us from the sun.

Scientists also now know that gravitational waves are real and measurable, adding to our general knowledge of waves and how the smallest and largest pieces of our universe interact.

It can be tough to wrap your mind around the concept of waves as things that exist but don't have mass. Instead of thinking about what they aren't, think more about what they are- integral, moving, delivery systems. Without waves, we would have no life-giving light from the sun and wouldn't be able to speak to each other. Waves are necessary and incredible. Now that we've learned the fundamentals behind particle and waves, we're going to look at a theory that ties together these two concepts and gives us a greater understanding of the big picture of physics by further examining the tiniest parts.

de Broglie's Hypothesis of Particles and Waves

French physicist Louis de Broglie (who was also the 7th Duc de Broglie) came to prominence in the mid to late 1920s for his pioneering work on wave-particle duality that we'll be discussing at length in this section. de Broglie grew up with a love of military history and rhetoric, and his first higher degree was in the humanities.

He would go on to study and receive degrees in mathematics and physics and was known as a prolific learner and exemplary student.

In 1914, de Broglie entered into the French army to serve in World War I.

During this service, he was stationed in Paris and assigned to develop, maintain, and operate radio transmission units, most famously, the one mounted on the Eiffel Tower.

He would also be among the first to help install radio communications equipment in submarines.

It was this experience with radio waves that ignited de Broglie's previous casual interest in wave movement and behavior. When he was released from the army in 1919, de Broglie would begin conducting wave and particle experiments in his brother Maurice (also a physicist) laboratory. In 1924, he released his seminal work on the subject, Recherches sur la théorie des quanta, or Research on the Theory of the Quanta.

His theory stated that *"any moving particle or object has an associated wave."*.

He based his hypothesis on his studies of the work of Planck and Einstein, who had been conducting extensive research into the properties of light energy and wave-particle duality.

The wave-particle duality theory had been developed to explain the behavior of objects on a quantum scale.

de Broglie was interested in taking the wave-particle theory down to the quantum level to decipher the activity he was seeing in electrons.

He suspected that they were behaving much in the same way as Einstein had proven light to behave when he theorized the existence of photons.

Certain that electrons were also acting as and traveling in waves, de Broglie and his colleagues set about finding a way to prove the hypothesis.

The resulting research led the scientists to the subject of our next section, de Broglie's Equation.

de Broglie's Equation

The de Broglie equation is an adaptation of some of Planck and Einstein's earlier equations explaining the behavior of light in the form of photons. Using work conducted by George Paget Thomson on diffracted cathode rays and experiments on electron behavior now known as the Davisson-Germer studies, de Broglie was able to conclude that particles can and do, in fact, act as waves.

This is the equation he created to explain and calculate that behavior:

$$\lambda = h/mv$$

In this equation, the Greek symbol lambda represents the wavelength, h is Planck's constant (of which we will discuss the development in a later chapter), *m* is the mass of the moving particle, and *v* is representative of the particle's velocity.

de Broglie's equation is used to prove that particles exhibit the same particle-wave duality as light.

The equation also serves to show that wavelengths change over time a distance, as their initial energy shifts from potential to kinetic and back to potential.

Have you ever seen a rhythmic gymnastics performance?

The gymnasts often use large, long ribbons to create beautiful visual effects while they complete their routines. But these ribbons can illustrate the loss of energy over the life of a wave, even without de Broglie's famous equation. You can recreate this experiment at home with a length of ribbon.

Taking your ribbon, hold it by the end in one hand, horizontally to the ground. By doing this, you are creating the plane along which your waves will travel. Now move your hand up and down in a fluid motion to create the amplitude of your wave. You will notice that the waves are more frequent nearer to the source of their energy (your hand) then they are at the end of the ribbon. This is because they lose energy over time, and the wavelength begins to increase.

The amplitude will also begin to decrease.

You've successfully demonstrated why and how de Broglie's equation calculates the average wave motion of a particle over time.

For his work, de Broglie was awarded the Nobel Prize in Physics in 1929, and Davisson and Germer would also receive the award in 1937 for their ability to prove the hypothesis in their laboratories. de Broglie went on to pioneer, extend, and test hypotheses on neutrino mass, thermodynamics, and duality in the laws of physics and nature, but it is his equation for which remains the best known.

Chapter 3
The Bohr Model, the Schrödinger Equation, and the impact of de Broglie's Hypothesis

One of the most incredible things about science is the scientific method itself. If you remember any of your junior high science classes, you'll surely remember learning that the root of all science is being able to produce documented, measurable, repeatable, tangible results. Maybe you had a teacher who was a stickler for properly kept lab notebooks.

For the men and women who revolutionized the field of quantum physics, carefully notating and documenting all their research allowed them to create detailed papers and books, giving their knowledge to the world.

It also allowed others to attempt to recreate their experiments to either prove or disprove their colleagues' theories.

In the case of quantum physics, the discipline erupted

and advanced so quickly during the late 19th and early 20th centuries that it wasn't long before everyone involved in the science was either building upon, modifying, or downright disproving everyone else's theories. One of the things that has stood the test of time, however, is de Broglie's hypothesis and de Broglie's equation. It did have a hiccup, though, when the model of the atom shifted from the Bohr model to the more accurate and advanced model proposed by Erwin Schrödinger in 1926.

Rethinking Atomic Structure and Particle-Wave Duality

Through the work of Planck, Einstein, and de Broglie, theories about the true nature of particles were being created, proven, and adapted to new research at an incredible rate during the first few decades of the 1900s. At the same time, other physicists continued to refine and recreate the working model of the atom.

While these two branches of research were happening independently, the impact and intertwining of this work are undeniable. When Planck and Einstein were embroiled their work on photon theory and particle-wave duality, the Bohr model of the atom was the one that was generally accepted.

As de Broglie worked to develop and prove his hypothesis to extend particle-wave duality to include electrons and other subatomic particles, Schrödinger released his updated model of the atom.

While the Rutherford and Bohr models were, and remain, great examples for teaching the basic structure of an atom, the Schrödinger model is a more accurate depiction of the behavior of the subatomic particles.

This model is not the two-dimensional graphic we think of when we imagine the models we first learned about in school.

The Schrödinger model is a three-dimensional view of the atom that gives scientists a more detailed concept of what the electrons of an atom are doing at any given time and also gave rise to Schrödinger's equation.

This mathematical sentence is what gave the Austrian physicist the nickname *"the father of quantum mechanics"*.

Before we get ahead of ourselves, let's make the connection between Schrödinger's equation, the de Broglie hypothesis, and their combined impact on the world of quantum physics. Schrödinger's equation, which is written as:

$$E\psi = H\psi$$

and is very similar in function to de Broglie's equation,

$$\lambda = h/mv.$$

Because this is quantum physics for beginners, and Schrödinger's equation is NOT for beginners, we've used the simplest form here and will give it the most basic definition.

The most important takeaway we want you to have

been the relationship that this equation has with the de Broglie equation and how they work together to form the basis of quantum mechanics and quantum physics.

Schrödinger's equation is used as a predictor.

The left side of the equation shows the available energy (E) in a closed wave system and the wave function (represented by the Greek letter Psi).

This indicates a prediction of where a particle could be at any given time in its wave movement.

It is shown in the equation as being equal to the same wave function and the Hamiltonian operator (H), a number that indicates the total of the potential and kinetic energy within the system.

It seems as if both sides of the equation are equal because they've just said the same thing in two different ways, but that's not exactly the case. Remember that the wave function itself (the variable indicated by Psi) is the result of a complicated derivative equation.

Its presence in this simplified linear equation is only because it's already been calculated and canceled out.

Schrödinger developed this equation because he wanted an easier way to impart the potential of a particle to move along a wavelength. He also wanted to prove that he could predict where a particle would be at any given time. Remember, Schrödinger is also known for his famous theory on a cat in a closed system- where he posited that a cat in a box could potentially be dead or alive and had an equal chance of being in either state. Still, no one could be sure until the box was opened, and the cat was observed. It was this type of philosophy that the eccentric scientist injected into his wave prediction equation. Schrödinger wanted to be able to find a way to convert the probability that a particle will be in a specific place into a linear equation that would represent that behavior.

If you are uncertain of any part of the equation, you can, much like Avogadro's equation that we discussed in Chapter 1, plug in any of the variables to determine the numbers you are lacking.

Relating Schrödinger's Equation to de Broglie's Equation

Schrödinger's equation may be complicated, but de Broglie's, thankfully, is not. When the first equation was published in 1924, de Broglie took a look at it and thought it was terrific, but he needed something easier to be able to perform his experiments and calculate the numbers that were going to help him in his research. Schrödinger's equation helped de Broglie gain a better understanding of wave function, ultimately leading to being able to construct his equation for wavelength. Here's the thing, though. de Broglie has a second, lesser-known equation, and once you learn that second equation, it all starts to come together.

$$\lambda = h/mv$$

To recap, Schrödinger's equation shows us the predictability of wave function, and de Broglie's equation shows us how to calculate wavelength based on the momentum (mass times velocity) of a particle. So, *what does de Broglie's second equation tell us, and how does it further relate to this early study of quantum mechanics?*

de Broglie's second equation is written as follows:

$$f = E/h$$

This equation shows that the frequency (f) of a particle wave is equal to its energy (E) divided by Planck's Constant (h). You probably feel like we're working backward since we haven't talked about the origins of Planck's Constant yet, but we will. It will be easier to understand its full meaning and impact if you first realize what a large role it plays in the equations that are most used in quantum physics and quantum mechanics. So, with two simple equations derived from Schrödinger's complex equation, de Broglie is able to explain both the behavior of wavelength over time and the frequency of waves given their energy levels. Fantastic! So *how did the evolution of the model of the atom from the linear, two-dimensional Bohr model to the more advanced model, proposed by Schrödinger himself, affect the way that physicists handled duality and wave function going forward?*

Adapting the Fundamentals as Knowledge Evolved

Up until 1926, most physicists developed and carried out their experiments using the Bohr model of the atom, which, as you'll recall, showed electrons traveling in fixed orbits around the nucleus.

Once Erwin Schrödinger proposed his equation for predicting where electrons could be based on their potential movement, Louis de Broglie was able to follow up with his equation to determine the wavelength of particles. It soon became apparent that electrons were likely not following the neat, circular routes depicted in Bohr's model.

Schrödinger proposed a new working model of the atom in 1926, and it quickly became widely accepted as the *"quantum mechanical model."*

This model is still in use today. The reason this model is considered to be more accurate is that the Bohr model is primarily two-dimensional. It is arranged in so-called valence shells, with the electrons more likely to be excited and peel off traveling in the outer orbits, and the more stable electrons are shown as traveling nearer to the nucleus. The Bohr model remains a good model for teaching the basics of atomic chemistry and physics to young students, but the physicists who worked and continue to work to advance the understanding of quantum behavior needed something closer to a true, three-dimensional working model of the atom. Enter Schrödinger's model.

Schrödinger recognized that the electrons are not only in constant motion but are behaving as waves rather than as particles. His model, the quantum mechanical model, reflects that evolution in understanding particle behavior. Schrödinger felt that his model would more accurately represent the constant fluctuation that occurs within an electron's orbit as it is pushed and pulled by the gravitational force of the nucleus. Rather than just the electrons in the outer valence shell of Bohr's model, any electron that happened to be near the outer range of the gravitational field would be the ones most likely to slough off or form connections with other atoms to make molecules.

Today, scientists use the quantum mechanical model of the atom as a basis for their experimentation.

They will often also use the term "probability cloud" to describe what they see in terms of the location of an atom's electrons. The distinct advantage of the Schrödinger model is that it is based on mathematical equations that can be calculated to show what an atom should be doing, even if the movement cannot be observed. The downside to this model is that even when a scientist has the ability to observe atomic behavior, wave movement on the particle level is still almost imperceptible. Even though researchers know this model to be mathematically sound, they may still lack the capability to observe it in action and prove it to be true. This is why some scientists are concerned that this model doesn't satisfy the Heisenberg uncertainty principle, while others hold that it does.

We'll be discussing that principal later in the book, so you can consider the evidence and judge for yourself. For now, let's move on to history, definition, and practical applications of the number you've been waiting for, Planck's Constant.

Chapter 4

The Planck Constant

We've already mentioned the Planck Constant or Planck's Constant in several contexts within the first few chapters of this book. It factors into many equations that are used frequently within the quantum physics community, so it's time to take a close look at its creator Max Planck, and why his work was so crucial to the advancement of the understanding of particle-wave duality.

Max Planck and His Early Work

Max Planck was a German physicist who came from a large family of scholars and academics. He received much of his grade-school level education in Munich, where he excelled in mathematics and mechanics, and was also known to be musically talented, training to sing and play multiple instruments. By all accounts, he could have pursued a career in classical performance but instead chose to follow his dream of being a physicist. By the early 1880s, Planck was considered to be one of the brightest rising young stars in the field, and by the end of that decade, he had already climbed his way up the ladder of academia to take a post at the Friedrich-Wilhelms-Universität in Berlin. When he retired from this position in 1926, he was succeeded by none other than Erwin Schrödinger.

Planck was fascinated by thermodynamics, and much of his early research, including that for his first doctoral degree, focused on this study. He was also interested in entropy, a concept which he felt "spooked" many of his colleagues.

His papers provided the basis for many others to begin proving their own theories, such as that of Svante Arrhenius's hypothesis of electrolytic dissolution. Planck would also become a much sought after lecturer, packing halls of interested students, many of whom praised him as the best speaker they'd ever heard.

Planck's many professional accomplishments, including winning the 1918 Nobel Prize for Physics for his discovery of energy quanta, were achieved over the course of a lifetime of personal loss.

War defined many of the moments of the physicist's life, beginning with the Prussian conflicts as a child and culminating in the tragic loss of many of his personal papers and research during the bombing of Berlin in WWII. He lost a son at the Battle of Verdun in WWI, another son was hanged as a traitor by the Nazis in 1945, and both his daughters died in childbirth.

He was widowed when he lost his first wife Marie to tuberculosis in 1918. He remarried and was survived by only his youngest child, a son named Hermann, and his second wife, Marga. During these times of personal and professional turmoil, Planck remained ever the stoic

German, refusing to turn his back on his Jewish colleagues during the rise of the Third Reich and throughout World War II. As the head of Germany's most prominent scientific societies, he adopted the motto of "persevere and keep working" and encouraged his contemporaries to do the same. He would continue to lecture until he neared his death in 1947, but his legacy in the field of quantum physics continues to this day.

Black Bodies and the Electromagnetic Spectrum

Planck's work and the development of the constant stemmed from his research into the electromagnetic spectrum and his theory on the behavior of black bodies. A black body is matter which collects and absorbs every particle of radiation with which it comes in contact. Furthermore, Planck hypothesized that the body could not only absorb all that radiation, but it could, in turn, store it and re-radiate it later.

Think of having a white cat and a pair of black pants.

If the cat sleeps on those pants, they will attract and hold onto most, if not all, of the fur that the cat sheds onto them. *When you shake the pants, what happens?*

The fur begins to fly back from the fabric and into the surrounding atmosphere. Planck wanted to know if a black body, in a vacuum, would collect, absorb, and then radiate all the energy it encountered, or if it needed to be acted upon by an outside force for this to happen. He was also curious to know what would happen in an open system, such as the cat and the pants. Black-body radiation is dependent upon thermodynamics and thermostability.

For this reason, it is also sometimes called thermal radiation or temperature radiation.

On a large scale, the best example of black-body radiation is a black hole, which absorbs everything within a radius commensurate with its mass.

As it absorbs more mass and energy, it begins to grow and increase its radius or *"event horizon"*, increasing its gravitational pull. Because it absorbs all electromagnetic waves, including the visible light spectrum, the *"hole"* appears black.

Remember, a black hole isn't a literal hole, but an object that's mass is so dense, it appears as a colorless singularity. Planck, and many others, theorized that a black hole retains the energy and mass that it collects, but that, like all things, would reach a breaking point and begin radiating all that energy back outward.

He hypothesized that any change in temperature or massive energy fluctuation would throw off the system the black body was operating in and force reversal of the absorption; in other words, the black body will begin radiating all of the energy that it previously took in. Much later, famed physicist Dr. Stephen Hawking would hypothesize that black holes and black bodies are always re-radiating absorbed energy, based on thermodynamic changes found along the event horizons of known black holes. There are, of course, also theoretical physicists who believe that singularities such as black holes could be the key to time travel, but we'll talk about that a little bit in our chapter on Einstein.

Planck's Law and Development of the Constant

So, you may be wondering what black holes, which are massive, have to do with quantum physics, which deals with microscopic particles. Planck was focused on finding an explanation for the behavior of visible light and the temperature at which radiation absorbed by and radiation emitted by a black body attain equilibrium. For example, the sun can be considered a black body, although imperfect, because it both contains enough mass to gravitationally draw in radiation from the area around it and emit radiation back in the form of light and heat. The temperature at which the sun reaches equilibrium is 5,777 degrees Kelvin (9938° F, 5503°C). This number is also referred to as the *"effective temperature"*. The effective temperature varies by the black body.

Planck had been searching for a way to work through a problem known as the *"Ultraviolet Catastrophe"*, which we can all agree is a rather dramatic name for a physics conundrum.

The ultraviolet catastrophe was an anomaly being

observed by physicists trying to explain the behavior of black bodies as they emitted radiation.

Many of Planck's contemporaries were observing this catastrophic event in their research.

While the scientists believed that a black-body should radiate energy at a steady rate across the broad electromagnetic spectrum, they were instead finding that the black-bodies were emitting large amounts of radiation in high-energy, high-frequency bursts, which would rapidly expend the absorbed energy and drop the system down to net-zero faster than expected. This effect was most often observed as the energy being radiated reached the ultraviolet range of the electromagnetic spectrum.

In his attempts to understand and resolve the ultraviolet catastrophe, Planck discovered that the problem with the classical physics being applied to the conundrum was that they didn't account for the full spectrum of electromagnetic radiation to drop in frequency and wavelength over time and change in temperature.

By adding these variables into the equation, Planck was able to develop Planck's Law.

It uses mathematics to describe the relationship between the energy absorbed by a black-body and the rate of release of that radiation at a certain temperature, considering that the rate of energy change could only be emitted in increments proportional to the spectral density of the electromagnetic wave.

In simplified terms, Planck's Law describes a closed system wherein the energy absorbed, and the energy radiated by a black-body under a constant temperature remain in equilibrium but accounts for changes in frequency and wavelength of the radiation given the potential energy and net-zero nature of the closed system.

When applied mathematics is used to prove Planck's Law, the results can be plotted on a curve that shows that the frequency of the electromagnetic waves will fall off after a certain time, given the type of radiation. Planck and his colleagues referred to this action as spectral density. The ability to express this behavior radically advanced the field of quantum physics and separated it even further from classical theorists.

Many scientists mark the publication of Planck's Law in

1901 as the "birth" of modern quantum physics.

Measurement and behavior
Planck's constant in action

One of the key factors in developing Planck's Law is the use of the number we're all here for, Planck's Constant. The constant is referenced in Planck's previous work but wouldn't be universally recognized as a mathematical constant until after 1905. There's a straightforward way to think about Planck's Constant, even before we get into any math. The basis behind the constant was Planck's desire to put a name or unit to the smallest possible amount of energy. That's all. Planck knew that the smallest pieces of matter had been discovered (at the time, this was the atom and its subatomic parts). He wanted a way to "quantize" or measure energy at its tiniest little wave. It was in his pursuit of this that Planck's Constant was born. Behold, the mathematical object of our affection:

$$h = 6.6262 \times 10^{-34} \text{ Joule} \cdot \text{second}$$

Let's break down what the numbers mean and how Planck came to them.

Frankly, the h is simply the variable letter that Planck chose because it wasn't being used to represent anything else in mathematics or the budding field of quantum physics. The SI unit joule-second is not to be confused with joules per second. A joule-second stands alone as a unit to measure both time and action.

Now for the number itself. 6.6262×10^{-34} is a really tiny number that represents the amount of energy that is produced by a single particle. We know that all particles vibrate.

Planck was the first to quantify or *"quantize"* that vibration.

The simplest way that Planck's Constant is used is to determine the energy of a photon by multiplying the constant by the photon's frequency. This works because we know that a particle's, such as a photon's, mass is equal to its energy. No matter which variables you possess, you will be able to calculate the ones you are missing, and all because of Planck's Constant. For example:

$$E = h f$$

In this standard equation showing the use of Planck's

Constant, we see that the energy (E) of a photon or particle is equal to the frequency (f) times the constant. This equation was developed as part of the Planck-Einstein relation and is a fundamental principle of quantum physics and quantum mechanics.

It was this equation that de Broglie took on step further in creating his own, which, as we know, calculates the behavior of a wave based on its momentum. Planck's Constant also plays heavily into the Heisenberg Uncertainty Principle, which we'll explore in some depth in the next chapter.

Development and use of Planck's reduced Constant another use for Planck

Another use for Planck's Constant is in its reduced form, symbolized by the h-bar, which looks like this: $\hbar$. The h-bar is used in place of the standard h in calculations that are factoring in angular momentum rather than linear momentum. Linear momentum is, of course, calculated by multiplying mass times velocity.

It depicts the momentum of an object or particle as it travels along two planes, most often in a straight line. Angular momentum is a product of calculating momentum in three dimensions.

A common example of this would be a gyroscope, which has the capability of moving in several directions and maintains its movement through the ability to adjust to those dimensions.

In classical physics, angular momentum is calculated through the sum of the momentum of all the moving parts, but this doesn't always work on a quantum scale. In order for physicists to be able to accurately determine the momentum of particles in three dimensions on a quantum scale, a new equation was needed. By using the Planck Constant in its standard form, physicists can use the de Broglie equation to solve for unknown momentums. For solving for unknown momentums in the case of a particle exhibiting angular momentum, a derivative of the Planck Constant was created, which we now call Planck's Reduced Constant, and the ℏ represents this new value. It is determined in equation form like so:

$$\hbar = \frac{h}{2\pi}$$

As you can see, the Planck Constant divided by two times pi gives us the reduced Planck Constant.

Why does this work to find unknown variables in problems involving particles moving in three dimensions?

To understand this, you must also understand that a wave is part of a parabola.

We know that if a parabola is extrapolated past its curve, it can eventually connect and form a full circle or 360°.

However, waves don't naturally turn back on

themselves and complete a 360° circle.

Instead, scientists measure one full wave cycle, from its starting plane (the baseline) up to the top of its amplitude (the crest) and back down through the baseline to its lowest point (the trough) as 360°.

Each time the wave completes this motion is measured as one hertz, and this is considered the frequency of the wave.

This is not to be confused with wavelength, which measures the distance between the crests of a wave.

By dividing the Planck Constant by 2π (the standard equation for determining the circumference of a 360* circle or wave frequency), the reduced constant can be used to calculate the momentum of objects or particles that are moving along more than one plane at a time. An example of adapting de Broglie's equation to use the reduced Planck Constant is:

$$\mathbf{p} = \hbar\, k$$

In this example, the variable **p** stands for momentum, the **h**-bar shows Planck's Reduced Constant (calculated by using the frequency of the wave in question), and the **k** represents the angular wavenumber.

Angular wavenumber is an overstated term for the measurement of waves occurring over a certain distance, rather than measuring them in time.

While Planck's Reduced Constant isn't used nearly as much as the standard constant, it is useful in determining the movement and momentum in those cases where a particle or object is traveling along more than two planes.

Planck himself was often nonchalant about his work and would frequently tell people, as in the case of the constant, that he was just looking for numbers that would make other numbers make sense.

He even once referred to the constant as a *"math trick."* He was a brilliant mind who probably discounted most of his own research as a means to an end, and Planck would likely be surprised about the impact of his legacy in the future development of quantum physics. But, truth be told, without Planck, his laws, and his constant, man may have never achieved space travel or built advanced research machinery such as the Large Hadron Collider.

Chapter 5

Heisenberg's Uncertainty Principle

We've all been uncertain at times.

Do we want the chicken or the fish? Which movie do we want to see?

Eventually, you make a decision, and the uncertainty is gone. But in order to understand the next concept we're about to tackle, you need to think about being both certain, and uncertain at the same time. Heisenberg's Uncertainty Principle, which he introduced to the world in 1927, aims to explain one of quantum mechanics' biggest problems, *how can one predict where a particle will be at any given time, even with the knowledge of its momentum or previous position?* First, let's look at Heisenberg's work that led him up to the Uncertainty Principle.

Heisenberg's Beginnings in Physics

Werner Heisenberg was born in Germany to academic parents. His father was a professor of ancient languages and Greek philosophy, and young Werner loved to engage in philosophical discussions with his own teachers and peers. He spoke almost lovingly of the atom as a philosophical pursuit, which could only be reliably accounted for with mathematics.

He would study under and with some of the other great scientific minds of his time, including Niels Bohr himself.

Heisenberg was also musically talented, a common thread among many of the pioneering physicists.

His propensity for the piano led him to meet his future wife, Elizabeth, after a performance. She was also from an academic family and encouraged him throughout his career to push his theories and research to new heights of discovery. The physicist was also an avid outdoorsman, active in many roles with the German Scouts throughout his lifetime.

He would often retreat to the mountains when he was

thinking through an incredibly difficult physics or mathematical problem.

While he is mainly known today for his famous uncertainty principle, Heisenberg's earliest major work was a collaboration borne from his doctoral thesis.

In partnership with Max Born and Pascual Jordan, Heisenberg proposed a set of mathematical matrices that could be used to describe and predict the motion of atomic particles in relation to mechanical processes. Unfortunately for Heisenberg and his colleagues, they were in the Bohr camp of theoretical physics, which was slowly being phased out for the more progressive work of Einstein, Planck, Schrödinger, de Broglie. While classical physics and mathematics were still a foundation of the newer fields of quantum physics, quantum mechanics, and atomic studies, the disciplines were experiencing a rapidly widening gap in beliefs and principles. While Heisenberg's mechanical matrices were not universally accepted or utilized by the physics community, they weren't without merit.

Part of the reason they fell by the wayside is that Bohr's school was falling out of favor as being outdated.

While this seems a little ridiculous given the speed at which new quantum discoveries were being made, Bohr and his contemporaries and students were firmly entrenched in the physical properties of the atom as a real, tangible object.

While the Einstein camp was studying wave-particle duality, the Bohr camp was concerned with what they called discrete bundles- quantum particles traveling together in packets of energy. They weren't interested in anything they couldn't measure through observation or predict with one hundred percent certainty.

While Heisenberg would move away from his previous colleagues in thought and action, due partly to Jordan leaving academia to become as Nazi SS officer in the 1930s, he would, later in life, give credit to Born and Jordan as being instrumental to his early development and eventual reception of the Nobel Prize.

Heisenberg himself would spend much of the 1930s and 1940s under the scrutiny of the Nazis.

They deemed his work to be counterproductive to their interest in harnessing nuclear power solely for the purpose of weaponization.

The Development of the Uncertainty Principle

The Heisenberg Uncertainty Principle has become Heisenberg's lasting legacy in the world of particle physics, and it was a long time in development and refinement. Heisenberg would never completely leave behind his belief in the Bohr school of study. Still, he would eventually have to recognize that the work of those in the Einstein school was garnering more attention. Heisenberg's views of the studies conducted by those who were working in conjunction with Einstein were complicated. He saw their work as dealing in *"reality"* and considered himself an *"antirealist"*. The contradiction here is that Heisenberg loved the mathematics of physics, which deals mostly in reality. Numbers are absolutes and are very real.

So, *where did the Uncertainty Principle stem from?*

Let's look at the basic premise of the Uncertainty Principle (which, by the way, Heisenberg himself, called the Indeterminacy Principle).

It states that it is impossible to know both the position and the momentum of a particle at the same time, even using observation, predictors, and equations.

That's a pretty, bold statement, so let's look at why it is both true and controversial. Obviously, if you can see and measure something, then surely it is exactly what you think it is and where you expect it to be.

To this day, scientists argue this point. Many feel that the measuring of particles with precision is the only way to be sure of their behavior.

Why would Heisenberg be so uncertain about this?

Heisenberg postulated that the nature of quantum movement is that there is a limit to how much knowledge one can gain from it. He believed that there were forces at work within a quantum system that were beyond the scope of human observation and understanding. Heisenberg went so far as to theorize that the more accurately one variable within a system could be measured, the greater the inaccuracy of another measurement. In simple terms, the more precise the measurement of the position of a particle, the less precise the measurement of its momentum, and vice versa.

Why would he think this?

And more to the point, could he prove it with his beloved mathematics?

Using his previously presented mechanical matrices, Heisenberg set out to prove his indeterminacy principle, and sure enough, given the tiniest variations in particle movement and momentum, he proceeded to prove that **a**, times **b**, didn't always equal **b** time **a**.

The infinitesimal differences he observed in quantum movement served as the mathematical basis for what would become the Uncertainty Principle.

Remember, Heisenberg and his colleagues were concerned mostly with mechanical systems, meaning that the particles were not existing in a vacuum, as with electromagnetic systems. Heisenberg concluded that the existence of even the most minuscule outside forces were causing the atoms to behave in a way that made observation and measurement limited in scope, thereby limiting the knowledge one could gain from studying the system.

Because he was also a man of philosophy and action, Heisenberg also conducted what his cohorts deemed a *"thought experiment"*, although Niels Bohr would later admit that the scientific basis of the research was sound. To perform this experiment, Heisenberg tried to examine the behavior of atomic particles, namely electrons, using a gamma-ray microscope.

While observing these particles, he noticed that the gamma radiation was acting against the natural movement of the particles.

It was essentially *"kicking"* the electrons around, not allowing him to get an accurate picture of what the particle should be doing in their natural state. To get a more precise reading on the behavior of the electrons, Heisenberg then applied a stronger, more accurate microscope to the particles.

What occurred was even greater unpredictability from the electrons, which were now being acted upon by energy from the stronger microscope.

Heisenberg ultimately posited that it was a limitation of the nature of quantum movement itself and not the scope or limitations of the observational equipment itself that created the uncertainty paradox.

Every time he applied an observation tool that emitted more energy, he injected that energy into the system and further increased the uncertainty. In the most basic of terms, it is impossible to know what a particle will do, even if you know where it is.

If you know what it's doing, you can't pinpoint its exact location. This principle is now one of the foundations of particle physics, quantum mechanics, quantum chemistry, and theoretical physics.

When scientists want to use Heisenberg's Uncertainty Principle in their work, they look at all the mitigating factors that could affect their measurements and observations, including the abilities and limitations of their laboratory equipment.

They also consider the accuracy of their baseline data, the confidence they have in their prior or preparatory work and the work of others, and the previously known uncertainty of similar experiments or materials.

By gathering and collating this data before beginning an experiment, physicists and chemists can determine the potential for variations and margins of error within their research.

Mathematics uncertainty and the Planck Constant in Action

When developing an equation to express the Uncertainty Principle in actionable terms, it was necessary for Heisenberg to employ Planck's Reduced Constant, or the h-bar, that we discussed in the last chapter. The simplest form of this equation is shown here:

$$\Delta x \, \Delta p_x \geq \frac{\hbar}{2}$$

This equation is a visual representation of the principle, and you can see Planck's Reduced Constant on the right side of the mathematical sentence.

It is divided by two because there are two variables on the left side of the equation.

On that left side, we see two Greek deltas, which are the uncertainties.

The Δ followed by the variable *x* represents the measurement of the position of a quantum particle, and the Δ followed by the variable *px*, which represents the measurement of the particle's momentum.

The Δ itself stands for the standard deviation.

When we put it all together, the entire equation reads, *"The standard deviation of the position times the standard deviation of the momentum is greater than or equal to half of Planck's Reduced Constant."*

Broken down like this, it's not hard to see what Heisenberg was getting at with his principle.

The standard deviation is the amount above or below a predicted or previously measured location where the particle can be expected to be found or its predicted or previously measured momentum.

This will vary by particle and condition, of course.

He predicted that these amounts multiplied by each other would always come out to an equal or larger number than the reduced constant divided by the number of variables. If the number were ever to come out less, that would mean that the position and momentum of the particles were predicted with one hundred percent accuracy before they even came into the field of operation, which is statistically highly, highly improbable.

Separating uncertainty from the observer effect

Within all of science, there is a puzzle known as the Observer Effect. The premise of this effect is simple; every time an observation is made via either human, mechanical, or digital means, the results of the observation are affected by the very act.

So, *does that mean that all research results are false?*

No, and the variation in results is so usually so minimal that they are almost undetectable.

However, these variations still exist.

They are not, though, to be confused with the Uncertainty Principle.

The Uncertainty Principle should be held in separate regard as the Observer Effect because the Observer Effect is present in nearly every aspect of life, on both a microscopic and macroscopic scale.

You cannot see in a dark room without acting upon it with a light source. You can't observe an atomic particle without some tool with which to perform your observations. Each time we try to observe something, we must act upon it with some outside force, which will then enact change on the system. There also seems to some misinformation or misconceptions that the observer is always human, and that human error or interference is the defining factor in the Observer Effect.

This isn't true, the Observer Effect also occurs in the case of mechanical, robotic, or digital observational tools. The problem is that it is impossible to conduct any research without research tools.

We'd never be able to learn anything!

Thankfully, the actual effects of observation are mostly harmless and can be calculated away with a margin of error. This is, of course, barring any truly catastrophic interaction, such as knocking over an entire experiment or some other freak occurrence.

Corrections, rebuttals, and adaptations of the Uncertainty Principle

A lot has happened in the scientific world since Heisenberg first introduced the Uncertainty Principle, and over the decades, it's had its fair share of controversy and adaptations.

There is a camp, albeit small of modern scientists who have refuted the Heisenberg Uncertainty Principle in its entirety.

There is also a larger contingent who feels that the principle's spirit should be upheld but that it needs some modifications or clarifications to remain suitable for use.

Since the Uncertainty Principle can only be applied to quantum-level study, many scientists feel the need to adapt it up to a macro level, but that's not mathematically possible. The principle only applies to quantum material because classical physics has the means to directly observe the position and momentum of objects without the use of equipment that will affect the variables. In quantum mechanics, it is necessary to use measurement and observation tools that will impact the system. This is one of the fundamental differences between classical and quantum physics and mechanics.

There is also a school of thought that completely dismisses the Uncertainty Principle and solely embraces Schrödinger's wave equation, although this isn't the fairest approach either, because quantum physics and quantum mechanics deal with two sides of the same coin of atomic predictability and measurability.

The Uncertainty Principle offers more flexibility in the interpretation of data, which is ironic given Heisenberg's devotion to the absolutes of mathematics but not surprising, given his equal love of philosophy and rhetoric.

No matter which school of thought you fall into, the Uncertainty Principle was and will remain a landmark in the development of particle physics and quantum mechanics. There are many who reject it solely because they don't want to think about the possibility of being unable to know everything about particle behavior even if we have the tools to observe and measure it.

While the world is populated with many brave people, fear of the unknown is a limiting behavior of humans and isn't likely to be overcome any time soon.

The last controversial thought we'll offer about the Uncertainty Principle is this, and it's in support of the tenet: if instrumentation has become more and more precise over time, *why does the principle still hold?*

There are many that believe it's only become more prescient, as modern instrumentation is stronger and more accurate than it has ever been.

It's up to you to decide how you feel about the Uncertainty Principle, but perhaps you could take a long hike in the mountains and think about it, much as Heisenberg himself would have.

Chapter 6

Einstein and his Foundational Physics

We've come to our last chapter, and yes, we saved the biggest name (and face) in quantum physics for the grand finale. Albert Einstein was not only one of science's most brilliant contributors, but he was also one of the most prolific. Einstein is known the world over for being a pioneer in the study and understanding of quantum physics. Let's take a detailed look at the man himself and some of his most important, lasting accomplishments and contributions to the field of physics.

Early Life and Work

Albert Einstein was born in the Kingdom of Württemberg, a state of the German Empire, in 1879. Although his family was non-practicing Jewish, he would attend a Catholic school for his early childhood education. He spent much of that childhood in Munich, where his father and uncle built and ran an electrical supply company. Einstein excelled in math and science, writing notable papers on matter states before the age of 16. At the encouragement of his uncle, he began teaching himself Euclidean geometry and algebra and studying music and philosophy in his early teens. Young Albert surpassed every tutor his family could provide for him and was admitted to university at the Swiss Federal Technical School on his second try at the entrance exams. He'd failed the first due to a lack of general education.

With his father's permission, Einstein renounced his German citizenship and became a Swiss citizen to avoid mandatory military service. He would graduate from the technical school with top marks but would be frustrated at the lack of teaching positions in his field. Unable to find an academic job, the would-be scientist took a job in Switzerland's federal patent office- a choice which would change the course of science history.

While working at the patent office, Einstein reviewed a number of applications for inventions that claimed to harness electrical signals.

It was these patent applications that would jumpstart Einstein's fascination with the connection between charged particles and the nature of traveling atoms and light.

Not content to waste away in a low-level government job, Einstein worked toward his advanced degrees and discussed science and philosophy with his friends. He was finally awarded his doctorate from the University of Zurich in 1905, kicking off what has been called his *"miracle year"*, and to be frank, what he accomplished in 1905 alone is going to take up most of the rest of the chapter.

Not only did he present and defend his thesis on the determination of molecular dimensions, but he also published major papers on Brownian motion, the photoelectric effect, the theory of special relativity, and the mass equivalency equation, which is now known as the most famous equation in the world. It should be noted; this was the year that Einstein turned just 26 years old.

Brownian Motion

Einstein's first breakthrough paper of 1905 was his treatise on Brownian motion. In the simplest of terms, Brownian motion is the random movement of particles when suspended in a gas or fluid. This phenomenon is so-called for a botanist named Robert Brown who, in 1827, observed the movement of pollen suspended in water. Einstein was the first to lend any serious credence to Brown's notes of the event.

When Einstein published his paper, he did so to lend his support that Brownian motion was the result of the presence of atoms and molecules in the water, providing a conduit for the particles of pollen to move. Because the energy of each individual molecule means that there is no constant to the force being placed upon the suspended particles, their movement is observed as being random.

Taking this theory, a little bit further is the idea that none of the bombarded and bounced-about particles can be counted due to their randomness, nor can the atoms or molecules that make up the energetic medium.

However, Einstein did create equations to accompany his theories, although both equations (each about a page long) were replaced with simplified versions created by other scientists.

Despite his equations being phased out, and despite the theory of Brownian motion being proven in 1909 by Jean Perrin, rather than by Einstein himself, the original credit for the concept remains with Einstein.

In true Einstein fashion, he was more amused by the initial snub of his theory than he was proud of it when it was finally proven.

Einstein's take on Brownian motion would prove instrumental in the development of several other theories by both classical and theoretical physicists and those who were getting involved in the budding fields of quantum physics and quantum mechanics. Not the least of these subsequent theories are the kinetic theory of heat, Stoke's law, and the ideal law of gas.

The Photoelectric Effect

The photoelectric effect is another one of Einstein's 1905 breakthrough hypotheses, and it would change the scientific world's view on the way light travels and is transmitted. The photoelectric effect was identified by Einstein as the emission of electrons from a material when it is hit with electromagnetic radiation in the light spectrum. In other words, when light (ultraviolet through infrared) touches a substance, it causes that substance to release electrons. The reason that Einstein's paper was so innovative is that it was in direct contradiction to the electromagnetic theory of classical physics. That model showed a predictable flow of electrons along an electric field created by the force and energy of the surrounding current.

In Einstein's model of photoelectricity, the electrons do not flow but are instead flung from their parent substance in a rather violent manner. Imagine you are standing in front of a wall made of sheetrock.

What would happen if you threw something at the wall?

Depending on the momentum of what you threw, it might go through the wall. It might glance off the wall and fly back at you.

Or it could hit the wall and lose all its momentum and tumble to the floor. No matter which of those three reactions occurs, one thing is certain, and that is that bits of drywall are going to be released from the wall when and from where your object hits it.

You can think of the wall as the test substance, the object you throw as a beam of light energy, and the drywall that flies off the wall as the electrons being released.

Einstein was certainly not the first to suggest the photoelectric effect, but he was the first to be taken seriously. As early as the 1860s, scientists were suggesting that light had the characteristics of both particles and waves but weren't sure how to prove it. In the late 1880s, Heinrich Hertz was able to produce electromagnetic radiation. Still, he couldn't explain why his results changed when he used ultraviolet rays as opposed to visible light or infrared. As we now know, it's because the shorter wavelengths of ultraviolet carry more kinetic energy and have greater momentum than the longer wavelengths of infrared radiation.

The next person to tackle the mystery of light energy was JJ Thomson, who we met early in this book as the progenitor of the plum pudding model of the atom.

As you'll recall, Thomson also was the first to identify electrons. It was a scientist named Philipp Lenard who would bridge the gap between Thomson and Einstein when Lenard conducted extensive research on finding the minimum threshold at which light would discharge electrons from other materials. He played around with increasing the intensity of his light sources but could never find an explanation for why the materials were behaving the way they did. Enter Einstein, who would be the one to make the connections between the behavior of light and its actual nature, which is that light, having no mass, must be made of pure energy and therefore is a wave. However, as particle-wave duality tells us, light must also have a structure to travel through space, and therefore, light is made up of particles which we now know to be photons.

Without Einstein solving the enigma of the

photoelectric effect, we would never have come to a full understanding of particle wave duality.

He was able to explain why Lenard's experiments with light intensity were not producing the expected results- it wasn't the amplitude of his waves that were lacking, but rather the frequency.

By increasing the frequency of the waves in the experimental process, Einstein was able to come up with the results that Lenard had been looking for, which was an increase in electrons released from a metal plate when hit with light waves.

With one paper on the photoelectric effect, Einstein turned the theoretical physics world on its head.

By postulating that light was, in fact, a stream of particles behaving as a wave, the face of quantum physics changed forever. In proper Einstein form, he, of course, was able to create a series of equations to quantify his theory, and unlike his equations for Brownian motion, these ones stuck.

To put math to work on the photoelectric effect, Einstein's equation looks like this:

$$\mathbf{K\,max} = h\,v - W$$

Look! There's our old friend Planck's Constant, showing up to lend a hand. Let's break down what's happening in this equation, beginning with the K on the left side. This variable, with its subscript, stands for the maximum kinetic energy of the electrons on the surface before being subjected to a light wave.

On the right side of the equation, we see the variables we'll need to determine that maximum kinetic energy. The h is Planck's Constant, and it's being multiplied by v, which is the frequency of the wave being applied to the electrons. The total of hv then has the last variable W subtracted, W being the work function of the electrons. This is the minimum energy threshold needed to remove electrons from the surface.

To understand the work, function a little bit better, it might be helpful to know that this variable is sometimes represented as BE, which stands for binding energy.

It is the work of the scientist to determine what the threshold frequency is for the waves they are using versus the material they wish to remove electrons from. The higher the frequency of the waves, the more likely the results will show more electrons being released.

This relationship will increase proportionately in between materials with robust and stable electron bonds these materials will need electromagnetic waves of increasing frequency to get them to shed their electrons. His work explaining the photoelectric effect was so seminal, it is the concept that won Einstein his Nobel Prize in 1921, despite the rest of his groundbreaking theories. By categorizing light as both a wave and a particle, Einstein opened the doors for the research and advancement of so many other theories and gave birth to a whole new world of quantum possibilities.

General relativity, Special relativity, and Mass Equivalency

To understand why Einstein's theories of relativity and the concept of mass equivalency were and remain so important, we must go back in time a little bit.

Because all of physics is built upon the work of the scientists who came before, we first must look at the two older components that go were the primary factors in what would become Einstein's hypotheses.

The first factor is the classical laws of motion, developed by Sir Isaac Newton in the late 1600s.

From Newton's world changing theories, we know the following things:

1) A body in motion stays in motion, and a body at rest stays at rest unless acted upon by an outside force.

2) Force equals the change in momentum per unit of change in time. In terms of constant mass, force equals mass times acceleration.

3) Every action has an equal and opposite reaction.

Newton's laws were the basis of classical physics and went unquestioned for nearly two centuries.

It's only once they began to be examined more closely that the deviation between classical physics and quantum physics began.

Quantum particles, as we know, do not behave in the same way macro-objects.

The second thing we need to factor into the background of the development of the theories of relativity is the discovery of the speed of light and early work into the nature of light.

A Scottish physicist named James Maxwell was the first to determine the speed of light (186,000 miles per second) in 1865, and he also suggested that light exhibits the properties of both a wave and a particle. However, Maxwell and his colleagues remained under the impression that light required a medium through which to travel.

In the 1880s, a pair of American scientists cracked the code on whether light needed a medium or if it could travel in a vacuum.

It sounds like the beginning of a joke, but a physicist and a chemist walked into a bar and bet each other that they could figure out the mysteries of light.

Okay, that's not exactly how it went, but the result is that AA Michelson and Edward Morley determined that light need no *"ether"* to surround it and can travel through space and time on its own, thank you very much. This revelation forever changed the way that scientists, and all of us, think about the very nature of life.

When Einstein was just a teenager in the 1890s, he was fascinated by the movement and nature of light, writing extensive papers, and performing his famous *"thought experiments"* about the subject.

He would write of one such experiment, where he pictured himself riding a wave of light and saw another wave of light running in parallel.

Despite his mass being atop the first wave, the speed was unaffected, and the two beams of light continued to travel at the same speed. What the young Einstein has stumbled upon was the origins of his theories of relativity.

Classical physics would have told Einstein that if he were atop a moving wave, running in parallel to another wave moving at the same speed, then the relative speed of the waves would be net zero.

This, however, is a direct contradiction to Maxwell's proven point that light always travels at the same speed, which we know to be 186,000 miles per second.

This got Einstein to thinking, *how can light beams traveling next to each other have both the same speed of 186,000 miles per second but also have a relative speed of zero?*

If you've followed along so far, here's what we can conclude: two objects moving at the exact same speed along the same axis will have the same point of view and see the same things.

It's all simultaneous, and their relative speed is zero. This follows the theories of classical physics, which Einstein did not disagree with.

However, if two objects aren't moving at the same speed, their relative speed is the difference between the two speeds. Imagine trains running on parallel tracks. One train is going 100 kph, and the other is going at 50 kph. These trains weigh the same, leave the station simultaneously, and reach their maximum acceleration at the same time, but the faster train reaches their destination in half the time as the slower train because they have a relative speed differential of 50 kph.

The first train travels 100 miles over the course of one hour; the second train takes two hours to achieve the same distance.

Light doesn't have this problem. Light always travels at the same speed and doesn't have to worry about resistance, friction, or other opposing forces.

If two beams of light replaced the trains in the previous example, those beams of light would reach their destination at the same time.

They always have a relative speed of zero. Now, let's add another variable to this. Going back to trains, let's say that a train is traveling past a fixed point, like a mile marker. If there is a man standing next to the mile marker when the train passes at 100 kph, he would see the train traveling past him and could observe the entire train. The train and the man have a relative speed of 100 kph because the train is moving, and the man is stationary.

Now, let's put the man on a train moving in the opposite direction on a parallel track.

This train is also moving at 100 kph. Both trains left

their destinations at the same time and achieved maximum acceleration simultaneously. They will pass the mile marker in the center of the route at the same time. This mile marker becomes a singularity, and now that both trains pass it, their relative speed will become zero, and it will seem as if time has slowed down.

This is the phenomenon that Einstein was most interested in. His curiosity about the simultaneous nature of movement in relation to time led directly to his creation of the special theory of relativity and the discovery of the space-time continuum.

Einstein wondered why, regardless of speed and position; objects could never outrun light.

He also wondered how time played into the equation.

The theory he created implies that the speed of light is the absolute limit of speed in the universe and that no object can ever overtake the speed of light due to the nature of mass. Because light has no mass, it is the only thing that can travel at that speed. He also theorized that mass increases as an object's speed increases and that eventually, the object will become so heavy that its mass becomes the limiting factor. That leads us to the most famous equation in the world, the mass equivalency equation, and it is brilliant in its simplicity:

$$E = mc^2$$

Let's break it down. The left side of the equation is where we see the variable E, which represents the total energy of an object. Since we know that mass and energy cannot be created or destroyed, we know that there is mass equivalency in objects that exhibit wave-particle duality.

This equation shows us what happens to an object that is traveling at the squared speed of light (notated here with the letter c).

This is a number that is microscopically shy of 90,000,000,000 square kilometers per second.

This number is then multiplied by the mass of the object, let's say 10 kilograms. The energy in that mass is now 900,000,000,000 joules.

That's a ridiculous amount of energy! But it's still not enough energy to move that object faster than the speed of light.

Einstein found that the faster an object goes, the heavier it gets. Its mass increases, and it is physically harder to move an object with more mass.

As the object hurtles along approaching the speed of light, its ever-increasing mass disallows it from ever reaching maximum speed. Therefore, the object will never be able to go faster than the speed of light, the fastest speed allowable in our universe. The simplest statement of the special theory of relativity is this: as an object approaches the speed of light, the closer to infinite its mass becomes, meaning it will never be able to overtake the speed of light.

Don't worry if special relativity seems counterintuitive to you; it felt that way to Einstein, too, and his contemporaries.

How can something continue to move that quickly and never achieve maximum speed?

It's because when we think about things moving, we tend to think about them having the ability to move in three dimensions up and down on a vertical axis, left and right on a horizontal axis, and forward and backward on a rotating axis.

But Einstein saw things a little bit differently and proposed that there is a fourth dimension that needs to be considered, and that fourth dimension is time.

Einstein posited that time MUST be accounted for when looking at relative motion and speeds. The idea of time as the fourth dimension had been tossed around by other physicists prior to Einstein's interest in relativity. He saw the work of German mathematician Hermann Minkowski, as a good basis for working out exactly how time factored into relativity. Minkowski published a paper in 1908 that solidified his mathematical theories on spacetime as the fourth dimension, and Einstein was fascinated with the matrices that included time as a vector that could be a fixed point, just as the standard points along the x, y, and z axes could be. This was the foundation that led Einstein to believe that time could be used as a coordinate, and so the concept of spacetime was born. Einstein also began to wonder what would happen if we stopped thinking about being in motion through space and started thinking about space moving around us.

Have you ever stood on the beach and let an ocean wave come in over your feet and legs? You are the fixed point, and the ocean is the body in motion. Yet, if you stand and gaze at another fixed point on the horizon when that wave crashes over your feet, you will feel as if you are moving backward when the wave recedes. We feel the same phenomenon sometimes when driving along a multi-lane highway. Because it is not statistically probable that every vehicle is traveling at precisely the same speed, there will be times when you look at the vehicle next to you and get the perception that the other car is moving backward rather than you are moving forward. These are everyday examples of relativity and the space-time continuum at work.

When Einstein began to think about time as the fourth dimension, he began to wonder why time seemed to slow down when objects were accelerating.

This train of thought is what brought Einstein from the theory of special relativity, which only involved objects traveling along a fixed plane at a fixed speed, to the theory of general relativity, which concerns all objects in space and time moving at a variety of speeds.

By adding acceleration and time as a dimension into his considerations, Einstein was able to come up with the following hypothesis.

Space and time are the two components of space-time, and the resultant forces within space-time (force, mass, acceleration) combine to create the phenomenon known as gravity. The gravity of objects in space-time has a warping effect on the space-time around them, causing an ongoing push-pull in the universe between objects of greater and lesser mass.

With this theory, Einstein had essentially solved one of the greatest mysteries of the universe. Prior to the release of his paper, "The Foundation of the General Theory of Relativity," in 1915, scientists understood what gravity was, but they didn't understand why gravity works. The easiest way to explain general relativity is to imagine that space-time is a giant sheet of fabric. If you place a large object in the center of the fabric, it will create a dip, and smaller objects placed on the fabric will begin to roll towards the larger object. This represents the largest object's gravity. But each of those smaller objects has its own mass and creates its own small dip. Whether or not those small objects roll all the way to meet the large object depends on how much gravity they each possess. Mass and gravity are directly related, the larger the mass, the stronger the gravity.

If the smaller object can create a dip large enough, it will prevent them from rolling all the way to the larger object's position and help them maintain their own place in space-time.

This is one of the reasons our solar system works.

The sun maintains the heaviest mass in the center of our system, and the planets all orbit around the sun, but each is also sitting in its own dip, preventing them from "rolling downhill" into the sun. There are also other forces at play that prevent everything from being drawn into the sun, such as individual rotation.

The counteraction of spin and gravity keeps each planet from leaving its orbit and being "sucked into" the sun's mass.

This also helps us explain the existence and behavior of black holes. While our sun has a tremendous amount of mass (1.989×10^{30} kg), black holes can have an average of three to ten times that mass. The black hole at the center of our galaxy, the Milky Way, has a mass that is 4.3 million times that of our sun.

This shows us why nothing, including light, can escape the gravity of a black hole. Their mass is simply too great compared to anything else that exists in the surrounding universe. As far as Einstein's view on whether or not singularities could signal the existence of wormholes for the use of time travel, the physicist felt that theoretically, this was possible.

However, he also believed that if nothing could survive being drawn into the singularity of a black hole, then the chances of a human being able to survive passing through a wormhole was very low if not zero. Einstein performed a lot of his famous thought experiments about the possibilities of time travel but was never able to postulate a theory that could be proven.

General relativity also explains the phenomenon of time dilation, which is something that occurs within gravitational fields. Time, as we know or consider it to be, was first measured thousands of years ago. Early systems of timekeeping were developed by the Sumerians in 3500 B.C.E., and ancient Egyptian, Roman, and Greek societies also had systems for marking time, chiefly through the use of sundials. In the following centuries, the invention and usage of the pendulum began to show that time was causally related to the daily rotation of the Earth. Eventually, the system which we recognize today was refined, that of 60 seconds to a minute, 60 minutes to an hour, and 24 hours to a day.

However, Einstein wanted to be able to show that time dilation is a real phenomenon, and he thought that his theory of general relativity would be a perfect conduit for this concept. Time dilation occurs when gravity affects not only the space around it but also the time. This can be proven with a simple experiment here on Earth; someone who is on top of a mountain and someone in the bottom of a valley can both be carrying the same exact timekeeping device, but time will move faster on top of the mountain. *Why is this?*

Because there is less gravity, the farther away you get from the center of Earth's mass.

The gravity in the valley is strong enough to literally slow down time.

This is the simplest on-world example of time dilation.

Now, on the Earth's surface, between the valley and the

mountaintop, the time dilation is perceptible but not monumental. However, if you put the concept of time dilation on a larger scale, such as the gravitational draw of a star or a black hole, you can begin to see why this is a major consideration in Einstein's theory of general relativity. The slowing of time near objects with a dense mass and gravitational pull has implications on all objects moving through space-time. It is a limiting factor on interstellar space travel, which we hope someday to achieve, and explains why we see the acceleration of objects towards objects with higher gravitational pull. That gravity is not only affecting the speed and acceleration of the objects that it is pulling in, but that gravitational pull is also making time itself move faster.

Another concept that Einstein felt was explained by general relativity is that of freefall.

We tend to think about falling in everyday life as a function of acceleration and gravity.

We know from our basic scientific knowledge that mass, velocity, time, and force can all be used to calculate how fast something will fall.

But Einstein wasn't interested in classical physics and mechanics. What he wanted to explore was the idea of falling without the opposing forces of friction, gravity, and resistance- in other words, freefall.

Using the theory of general relativity, Einstein was able to conclude that absent any forces of gravity, an object could theoretically fall forever until it was acted upon by an outside force or object, namely, landing upon a surface. Einstein hypothesized that since all objects in freefall experience the same acceleration regardless of mass when the gravitational force is the Earth-standard 9.8m/sec2, then when gravity was increased or decreased due to a warp or flattening of space-time (remember our fabric example?), then the acceleration or deceleration of freefall would also be affected.

General relativity offers a lot to chew on.

But by understanding just these few major tenets, it's easy to see why Einstein's theories had and still have such an impact on almost every breakthrough in quantum physics, quantum mechanics, and theoretical physics since.

Thinking about general relativity certainly gives you pause to consider your place in the universe, doesn't it?

From the tiniest particles and photons to the densest black holes, we occupy such a unique place to be able to study and comprehend both ends of the spectrum.

Later Years and Lasting Impacts of Einstein's Work

While it may seem like most Einstein's seminal works came while he was still young, the physicist enjoyed a long career teaching, traveling, and lecturing until he passed away in the United States in 1955. Like many of his theories, Einstein's life itself was complicated and marred by war and personal contradictions. One of the world's most brilliant scientific minds was not the best, even by his own admission, at interpersonal relationships. He had two marriages that were affected by his inability to stay faithful. He was also a renowned pacifist, a trait that would lend him a repeating theme of moral conflicts.

As you recall, Einstein renounced his birth citizenship in the German Empire to avoid mandatory military service, becoming a Swiss citizen in 1901. In 1914, he signed an official declaration that announced to all of Europe that he was a pacifist and a globalist, underlining his belief that his science belonged to all people, not just those of the nation in which he was living and working. At the conclusion of World War I, Einstein was invited to visit, tour, and lecture in the United States for the first time. He arrived in the US in 1921 and spent over a month using his appearances as a fundraiser for the Hebrew University in Jerusalem.

It was around 1926 when the two schools of thought in quantum physics and quantum mechanics, Einstein's school and Bohr's school, began to really diverge. In 1927, the two men would engage in a series of highly popular debates. This debate series began to open up the world of physics to the general public, and Einstein's pop culture stock began to soar once again. The rise of the Nazi party in Germany concerned Einstein, and so he accepted a position at Princeton University in New Jersey as the head of the Institute of Advanced Study. While he had intended to split his time between New Jersey and Berlin, his stance as a pacifist made him unwelcome in his native Germany. In 1933, he resigned from the Prussian Academy of Sciences and declared that he would likely never return to his homeland.

While working in the United States, Einstein was painfully aware that many other physicists were hard at work trying to harness nuclear power for use in weapons. He even signed off on a letter to President Franklin D.

Roosevelt explaining that German scientists were also working on the nuclear technology needed to create an atomic bomb. Although he was deeply rooted in his own pacifism, Einstein encouraged the United States president to make sure that the military was diligent in its own pursuit of nuclear weaponry. Einstein knew the significance of his work in the development of this technology and wanted to make sure the leader of his new home did as well.

Despite this, and despite becoming a United States citizen in 1940, Einstein would never work directly on the production of the atomic bomb. The scientists who were part of the Manhattan Project were expressly forbidden from talking to Einstein due to his left-wing political leanings. Although nuclear fission would not have been possible without Einstein's equation of mass equivalence, he would never work directly with atomic weaponry, a fact which didn't bother him in the slightest. He knew the impact he'd already had on its development, but because of his deep belief that science belongs to the people, Einstein also felt he had no control over what others chose to with his theories and findings.

The reason the **E=mc2**, was so vital in the advancement of nuclear technology is that it gave scientists a context within to work on splitting the atom. Nuclear fission is at the heart of atomic weaponry. Nuclear power itself is dependent on the natural decay of radioactivity, and soon the technology would be used to build power plants and nuclear-powered watercraft like submarines. Mass equivalency is what allowed scientists to recognize that they could use minimal amounts of dense radioactive elements, like uranium, to create large amounts of energy, like a nuclear explosion. In controlled settings like a power plant, radiation from highly radioactive elements is trapped as the materials decay and harnessed into electrical power.

In his later years, Einstein, who had always been considered a bit of an outlier, began to distance himself from the theories that his colleagues were presenting. He was unhappy with the direction that quantum mechanics was taking, and actively spoke out against his old contemporaries, even after his public debates with his friend Niels Bohr. As Bohr became more entrenched in mechanics, their opinions and schools of thought would never cross again. It was Heisenberg's declaration that the "quantum revolution" was over that sent Einstein firmly and with finality running from the establishment.

Although Einstein's major accomplishments all occurred before 1930, he never once stopped working on developing new theories, writing papers, and lecturing. He published hundreds of short works between 1930 and his death in 1955 and renewed his Jewish faith. He was even offered the presidency of the young Jewish sovereign state of Israel in 1952. At the time of his passing in 1955, he devoted all his time and research to developing what he called a "unified field theory," which he believed could be the master key to unlocking all of quantum, classical, and mechanical physics. While he wasn't successful in solidifying this hypothesis before his death, he would be happy to know he kickstarted a campaign that continues today to find the so-called "theory of everything."

Chapter 7
A glimpse into the future of Quantum study

Although the basic premises of quantum physics had been set by the mid-1900s, that doesn't mean that the advances made by the pioneers of the field stopped there. Scientists continue to work every day to unlock new and exciting theories about the behavior of atoms, particles, and waves. Our ever-increasing knowledge of these unseen things is what brings us some of the technology we enjoy in our everyday lives.

Think about the things you did when you got up this morning. Without the pioneers of electricity, you would have been able to turn on the light and start your coffeepot.

Without the scientists who discovered microwaves, you wouldn't be able to heat up your breakfast, and your cell phone wouldn't exist without electrical engineers and quantum physicists. Our lives would be completely different if we didn't have some of the most brilliant scientific minds furiously advancing the field of quantum study in the early part of the 20th century.

The great thing is, we have the same type of brilliant minds today who continue to advance our understanding of quantum behavior.

There are experiments going on worldwide in countless laboratories that are finding new things at a breakneck pace.

In the later part of the 2oth century, two significant branches of study began to emerge among those quantum scientists who wanted to continue to study the tiny things atoms, subatomic particles, photons, and quarks. These scientists and mathematicians are working steadily to identify and classify the behavior of the most infinitesimal parts of matter.

The second school of quantum physicists is those who want to apply what's known about the tiniest things to the largest things planets, stars, galaxies, and all the other bodies, seen and unseen, that make up the greater universe.

Astrophysicists would have nothing to study if it

weren't for particle physicists. Being able to predict the behavior of an atom means that we can change the behavior of everything that's made of atoms.

Similarly, a deep understanding of wave behavior means knowing what to expect from waves of all sizes, wavelengths, and forms of matter, including gravitational waves. Let's look at what's going on these days in both fields.

Quantum Mechanics in the 21st Century

There are a few areas of focus in the study of quantum mechanics in the 21st century.

The first and most well-known is the existence of the Higgs bosun, which was proven to be confirmed through experiments being run at the Large Hadron Collider at CERN in Switzerland in 2012.

This was the result of decades of study of the parts which make up particles. Over the course of the mid to late- 20th century, scientists were able to determine that even protons, electrons, and neutrons were made up of even smaller pieces, and the Higgs boson was one of the most elusive.

Another area of quantum mechanics that modern physicists are trying to work out is that of quantum entanglement. This is a conundrum that means that sometimes, the properties of one quantum substance cannot be distinguished from the properties of another because the substances or objects are *"entangled"* within a system. Despite knowing that there are two distinct objects with their own characteristics within the system, a scientist must observe the behavior of both materials in order to study one. The materials must be described in relation to each other, or the observations will be inaccurate.

One other central focus area of quantum mechanics is on quantum computing (using mathematics to predict the behavior of micro-quantum particles, like quarks and bosuns) and quantum transfers, which is the movement of data and matter using quantum-level communications. This is truly the stuff of science fiction because the result could eventually be the transfer of large matter particles using wave mechanics and quantum particle movement.

Of course, we are still years or decades away from being transported from place to place, as is depicted in popular sci-fi shows and movies, but it's tremendously fun to think about.

Of course, there is that whole pesky thing about not being able to be put back together properly, but all in good time.

Those who study quantum mechanics also continue to

expand their knowledge of wave behavior, which has a direct effect on everyday life in the 21st century. Broadband communications and increasingly speedy and reliable cellular phone networks are one fantastic benefit of the work of wave mechanic scientists. So too are the components that go into our communication and entertainment devices. Being able to build receivers and transponders that can handle rapidly evolving transmission equipment is equally important.

Quantum mechanics is also responsible for many of the other things we are all familiar with lasers, atomic clocks, computers, and MRI technology.

Even the technology that goes into things like satellite dishes and solar panels is all thanks to quantum mechanics and a fundamental understanding of how particles and waves work.

One of the most significant advancements in the last hundred years was the development of the electron microscope a gift to scientists from scientists.

As those who study quantum mechanics continue to increase their knowledge and understanding of the mechanical movement of the universe's tiniest particles, you can only imagine the advances that we will see over the next decades and centuries.

Quantum Physics in the 21st Century

Quantum physics is indelibly linked with quantum mechanics, but while some scientists choose to focus their energies on examining the infinitesimal particle that is the basis of all quantum studies, some choose to look at the big picture. Many quantum physicists these days spend time examining and re-examining Einstein's theories and apply them to be able to study the universe at large.

The development of space exploration and our deepening understanding of space is largely in part to Einstein's theories of relativity. Whether it is the technology that goes into massive telescopes that help us see the outer reaches of our solar system or galaxy or the inner workings of manned space flight, none of these things would be possible if Einstein didn't give the world the means by which to both put these objects and people in space and interpret the results of their studies.

Einstein's understanding of the behavior of matter and his explanation of the nature of gravity were instrumental in being able to learn about the universe beyond the confines of Earth. The theory of general relativity factors into nearly everything about space travel and exploration; it's also an excellent opportunity for scientists and astronauts to continue to prove these theories right. From being able to tell if radio transmission waves are bending to accommodate gravitational fields to being able to determine if planets are orbiting distant stars, the theory of general relativity is continually being used in space.

Quantum physics also played a huge role in being able to take the very first photograph of a black hole, which occurred in 2017. This was groundbreaking work for many obvious reasons, but it is also one of the greatest indicators that Einstein's theory literally holds true universally. The photo, taken over a period of five days using a series of eight telescopes in a world-wide collaboration, shows a massive gas cloud surrounding a black hole that is 54 light-years away from Earth.

A black hole is a familiar space object, but one that we may never know the exact nature of, and that's because its intense gravitational pull makes it nearly impossible to know what occurs "inside" the black hole itself. Modern quantum physicists and astrophysicists struggle to reconcile the knowledge they have, the knowledge they want, and the knowledge they may never gain. This doesn't stop these scientists from working tirelessly to understand the mysteries of the universe.

One major dilemma that faces quantum physicists is that ever since the schools of thought split between Bohr and Einstein, it has been challenging to come to grips with the fact that the two fundamentals of quantum studies- mechanics and relativity- are essentially at odds with each other. Both camps continue to find new explanations of how the universe works, but neither can fully agree with the other. Those who are working in the fields of relativity can, however, agree that there will always be new worlds and new matter to explore.

One such way that these scientists keep Einstein's

legacy alive is by trying to prove the existence of dark matter and dark energy. We know that matter and energy are equivalent, and we know that matter and energy cannot be created or destroyed. But there are unseen forces in the universe that can only be explained by the presence of energy and matter than we are yet to understand. So, *what are dark matter and dark energy, and what are scientists doing to try to understand it?*

In simple terms, dark matter is what remains after the known matter in the universe is accounted for.

This matter could be made up of black holes, brown dwarves, or other dense, colorless matter, although it's likely we would be able to see or detect the presence of such large or mass-dense objects. It's also theorized that dark matter is made up of the opposite of the particles that we're familiar with, although the theory of anti-matter is more likely still the purview of science fiction. It's most probable that dark matter, which makes up about 75-80% of the known universe, is a combination of yet-to-be-identified quantum particles, undetected black holes, and other dense neutron stars.

The most boring and most likely answer, though, is that

dark matter is made up of the same atoms and molecules as known matter; we just haven't been able to see it yet.

Dark energy is another story.

Dark energy is the force that seems to be causing the universe to be ever-expanding, and no one has quite figured it out yet. We know that dark energy exists. We know that it's behind the expansion of the universe, and we also know that it's causing that expansion to accelerate. What we don't know about dark energy is why it is doing this. For some time, astrophysicists were concerned that this rapid expansion could mean that the universe was working its way towards self-destruction- that like an elastic building up potential energy as it is stretched, it would eventually just snap back. This would cause the reverse of the Big Bang and has been dubbed the Big Crunch.

Now scientists view the continuing, quickening expansion of the universe as more of an infinite behavior. That even though we know that matter and energy cannot be created or destroyed, the universe will eventually have to wear itself out, and the expansion will either stop, or the universe will pop like a balloon. Thankfully, we won't have to think about any of these events occurring in our lifetime, but physicists are still working to find answers as to why dark matter and dark energy have such an effect on the known universe. The more answers they can gain, the better chance we all have of understanding the workings of all the matter we cannot see.

On a happier note, than the eventual destruction of the universe, quantum physicists are working in new ways to think about the phases of matter, and manned space flight gives them the opportunity to do so. Trained astronauts on the International Space Station have access to the vacuum of space to perform experiments on sublimation and condensation, as well as see how the space vacuum affects the ionization of a variety of elements.

Another benefit of being able to test theories and materials in the environs and vacuum of space is being privy to a zero-gravity environment.

Under these conditions, the forces of gravity cannot affect the particles that astronauts are studying.

Many of the people who now travel to the International Space Station to work are trained scientists who took on the additional burden of becoming astronauts, whereas, in decades past, it was scientists on Earth who would instruct astronauts on how to act as their research proxies while in space. The result is a cross-trained space contingent who are constantly testing the limits of the behavior of matter both native to space and introduced to that environment.

Scientists who work in observatories and at space agencies are always looking for new ways to gain more knowledge of the workings of the universe, including studies into the origins of matter, whether the speed of light really is the universal speed limit, and if there are new and exciting ways to apply Einstein's principles to "see" beyond the outer stretches of space by detecting new gravitational fields. No matter what the future holds, we can all rest assured that physicists are working hard to help us understand the very nature of the space we occupy, on both a tiny scale and a universal one.

We've now come near to the end of our time together, and it's been a heck of a journey through time to learn about physics!

Before you get to the conclusion of the book, you'll find two appendices, which are meant to help you recap and remember the concepts we covered. The first appendix is a timeline of early physics breakthroughs and discoveries, and the second is a list of formulas and equations that will be of use to you if you want to begin crunching numbers on your own.

There are so many different fields that erupted from the humble beginnings of classical physics.

If you're interested in how the world works, then you are interested in physics.

But with a panoply of choices, if you've decided that quantum physics isn't for you, then congrats! At least you made it all the way through the book before you decided that. Maybe you'd like to check out quantum mechanics, quantum statistics, or quantum electromagnetics. Perhaps you're more inclined to fall into the classical mechanics camp, where you could study thermodynamics, mechanical wave theory, or classical statistics.

If you like to dabble in the unknown, you may want to

explore the world of theoretical physics. You could hypothesize about black holes, string theory, wormholes, and time travel. The world needs more dreamers who are willing to back up their dreams with science. Many of the world's coolest and most loved inventions were created by scientists who dared to dream, so maybe you could be the next. No matter where you are in your life or where you want your science journey to take you, just remember that there's no wrong decision when you choose to study science.

Humans are innately curious beings, and our capability for higher thought and reasoning is what sets us apart from the rest of the animal kingdom. We could theorize, perform the scientific method, and gain answers through thought, action, words, and numbers. For all aspiring scientists, this is a comforting thought. Although atoms and particles are small, and the universe is vast, we can always take heart that science is concrete and won't lead us astray. We hope you enjoyed reading your way through the basics of quantum physics and that you've been inspired to take your scientific knowledge to the next level!

Appendix A: Timeline of Major Breakthroughs in Early Quantum Physics

Because the developments in the early study of quantum physics could often overlap or come at a rapid pace, here's a handy timeline of all the events and research we've discussed to help you keep track. It is by no means a complete timeline of every bit of physics research, but it is a guideline to help you remember the main points we outlined in the book:

1808 Dalton releases his hypothesis of the properties of the atom.

1865 Maxwell determines the speed of light.

1895 Röntgen discovers X-rays.

1897 Thomson discovers the electron.

1898 Becquerel discovers radioactivity; Curies begin their studies of radium/polonium.

1900 Planck quantizes particles after extensive research into black body radiation.

1903 Becquerel and the Curies win the Nobel Prize for

their work on radioactivity.

1904 Thomson releases the plum pudding model of the atom.

1905 Einstein proposes that light can also be quantized, introduces his photon theory.

- Einstein publishes his paper in support of Brownian motion.
- Einstein releases the theory of special relativity.
- Einstein presents the *"world's most famous equation"* **E=mc2.**

1909 The Geiger-Marsden experiments suggest the existence of an atomic center.

- Perrin coins the term "Avogadro's Constant" to describe the molar value.

1911 Rutherford uses the Geiger-Marsden research to propose the nucleus theory.

1913 Bohr presents his planetary model of the atom.

1915 Einstein officially presents the theory of general relativity.

1918 Planck wins the Nobel Prize for his Law and Constant.

1919 Rutherford discovers and names the proton.

1921 Einstein wins the Nobel Prize for his theory of the photoelectric effect.

- Chadwick theorizes the charge that holds an atom's nucleus together.

1923 Compton completes research that confirms the existence of photons.

1924 de Broglie generalizes the theory of wave-particle duality, introduces his equation.

1925 Bothe and Geiger apply the Laws of Conservation to atomic processes.

1926 Schrödinger introduces wave mechanics and equation.

- Lewis officially names the photon.

1927 Heisenberg presents his Uncertainty Principle.

1929 de Broglie wins the Nobel Prize for his work on wave-particle duality.

1932 Heisenberg wins the Nobel Prize for his introduction of quantum mechanics.

1933 Schrödinger wins the Nobel Prize for his creation of wave mechanics.

Appendix B: Formulas and Equations

A reference list of basic physics formulas and equations, as well as the quantum physics formulas discussed in this book.

- *Weight*

weight = mass times gravity **W = Mg**

- *Speed*

speed = distance/time **s = d/t**

also called *"velocity"* **v = d/t**

- *Acceleration*

acceleration = change in speed/time **a = (s1- s2)/t**

- *Force*

force = mass times acceleration **F = m a**

- *Momentum*

momentum = mass times velocity **p = m v**

- **Acceleration due to gravity**

where gravity is 9.8 m/s2, where g is the acceleration,

where m is mass, where r2 is the radius of the earth.

g = Gm/r2

- **Avogadro's number**

to determine molar content **6.02214 x 10^{23} = mol**

- **de Broglie's equation**

to prove particle-wave duality **λ = h/mv**

- **de Broglie's second equation**

to relate frequency to energy **f = E/h**

- **Schrödinger's equation**

to relate frequency and wavelength **$E^\psi = H^\psi$**

- **Planck's Constant**

to determine quantum energy **h = 6.6262 x 10^{-34} J·s**

and reduced constant **$\hbar = \dfrac{h}{2\pi}$**

- **Energy of a photon**

to determine the energy of a light particle **E = h f**

- **Heisenberg Uncertainty Principle**

to determine deviation $pq - qp = h/2\pi i$

- **Photoelectric effect**

to quantify potential for electron loss

$K_{max} = h\nu - W$

- **Mass equivalency**

to show limitations of mass (energy) and v. speed

$E = mc^2$

Conclusion

Thank you for reading Quantum Physics for Beginners. We laid out a tremendous amount of information in this book, and we hope that you've enjoyed this scientific journey with us. We also hope that this book has inspired you to further your study and delve deeper into the mysteries that quantum physics has to offer.

Within these chapters, we explored the very beginnings of quantum physics and spent some time giving you the background on the field's earliest trailblazers.

From Dalton's earliest description of the properties of the atom to Avogadro determining a way to quantify the number of atoms in a set quantity of material, quantum physics' earliest discoveries came quickly.

The work of Becquerel and Marie and Pierre Curie on radioactive materials gave the world surprising insight into the behavior of some of the world's most fascinating, useful, and dangerous elements.

The work of the earliest quantum physicists and chemists also led to medical knowledge that we continue to use and develop new, advanced versions of in the 21st century.

When scientists were able to begin using x-rays in a laboratory, it wasn't a far leap to start the practical use of them for medical diagnostics.

Marie Curie herself worked to outfit a fleet of ambulances with portable x-ray equipment on board for use by French troops in World War I.

Although the delivery has changed and evolved over the last century, the base technology of the x-ray machine has not.

Cancer treatments rely heavily on radiation and chemotherapy, all made possible by the early work of quantum physics and chemists.

The first 15 years of the 20th century brought a period of rapid change in the fields of quantum physics and quantum mechanics. The atom itself went through two different modeled depictions, first Thomson's plum pudding model, and then Bohr's working model. It would change again when Schrödinger introduced his model in the 1920s. But on top of learning about the structure of the atom, Einstein himself dominated the earliest part of the 1900s, releasing four major theories and his doctoral thesis in 1905 alone.

Schrödinger and de Broglie were the men who took over the 1920s, and their equations have stood the test of time. By refining our understanding of particle-wave duality, these two scientists were able to create a whole new school of thought and study, that of wave mechanics. Their equations were groundbreaking and continue to be an integral part of the field.

Without those who study wave mechanics, we wouldn't enjoy much of the personal electronics technology that we have in our homes today.

Whether or not you can wrap your brain around Einstein's theories of relativity, there is no denying that these hypotheses changed the world, not only in terms of scientific advancement but in everyday life and the course of history.

Who knows where we would be in terms of scientific development without the understanding of mass-energy equivalency?

We certainly might not have had the atomic bomb, but

we also might still be decades behind in the study of outer space. It stands to reason that with any major scientific breakthrough, there may be give and take when it is applied to everyday life and further scientific studies. Ethics plays a huge role in science, and physicists must always weigh their studies against the greater good.

The future of quantum physics is bright. Scientists continue to find and study smaller and smaller particles and get closer to defining the very tiniest building block of all matter. Astrophysicists are working to identify the true origins of all matter and explain its behavior throughout the universe. Knowing where we came from and how matter was formed at the very beginning of creation will help us appreciate and understand where we are headed in the coming centuries.

So, once again, thank you for reading Quantum Physics for Beginners. You've opened yourself up to the wonders of the seen and unseen universe, and we hope you decide to continue your studies.

The world needs as many curious, fearless science lovers as it can get.

www.ingramcontent.com/pod-product-compliance
Lightning Source LLC
Chambersburg PA
CBHW070639220526
45466CB00001B/229